This edition first publis…

Printed and bound in the UK

A CIP catalogue record for this book is available from the British library.

This book can be ordered direct from the publishers at:
sales@greentechpublishing.co.uk

www.greentechpublishing.com

ACKNOWLEDGEMENTS

I would like to thank all the people who have knowingly and unwittingly contributed throughout more than 20 years of my green transport career, by helping refine and enhance my knowledge and understanding of the electric car sector and its supporting infrastructure in general. These people, too numerous to mention, have helped shape the world of electric cars as we know them today and have ultimately assisted in my accumulated knowledge over the years, in producing this invaluable **Electric Cars: The Ultimate Guide**. Huge thanks must also be given, recognised and acknowledged, to my past clients and colleagues, many of whom are now lifelong friends and are still enthusiastically passing on their electric transport knowledge and benefits, to both the adopters and the unconverted.

I would also like to thank all the car manufacturers press and media departments for allowing me to use official photos and images in this book.

Finally, thanks go to the team at Greentech publishing who have provided the perfect balance of critical feedback and support, leading to this book's first edition in what will be an ongoing biennial update of **ELECTRIC CARS: The Ultimate Guide**.

To my friends, family, colleagues and fellow researchers, who for many years shared my vision of a global electric transport revolution.

The future is now.

CONTENTS

INTRODUCTION

My motivation for creating this guide for both potential and current EV users, was the sheer lack of unbiased information that is available on electric cars in this format. I trawled bookshops and the internet for months and could only locate highbrow academic papers focused on just a just a single facet of EV's, or at the opposite end of the scale, very poorly researched and produced manuals, that were so far out of date, that they were by now, consigned to history.

I have owned petrol, diesel and electric cars, and I can say with absolute conviction that the most satisfying to drive, the most economical, the most ecologically friendly and by far the best performing, have been electric cars. But I will let you be the judge after reading this guide. Ultimately, the first steps in this process is to gather as much information as possible, then draw up a shortlist, followed by a weekend test drive. Only then will you be able to make an informed decision.

I am an electrical engineer by profession and have worked on EV concept designing, prototyping, market testing, production and collaboration. Most recently I have been fortunate to be involved in pioneering EV charging infrastructure, including Europe's largest 20 bay multi-point battery supported rapid charging centre, based at Heathrow Airport in London. Having worked in the electric vehicle industry for more than 20 years, I decided to investigate the most up to date information available on electric cars to bring to market an informative guide encompassing the history, technology and buying and running an electric car.

The result is this guide, focused only on pure Battery Electric Cars, powered purely by an integrated battery pack, without any additional assistance or back up, such as Internal Combustion Engine (ICE) hybrids or ICE range extenders. The latter two are in my opinion not the future and are therefore not discussed.

This sector and its technology are changing at such a rapid rate, that I will be producing a new edition of this book at least every two years. The rate of change in the last 10 years has been incredible. We have gone from the rather pedestrian and low rent lead-acid battery powered G-Wizz, with a range of 40 miles and top speed of 50mph, to the current raft of EV's, some of which break the 350-mile real-world range per charge and top out at more than 140mph. By the time you have read this book, I hope that you will appreciate what an electric car is, how it works, the pros and cons, how far we have come technically and how far we are powering into the future. If you decide to proceed and purchase an electric car, you will have a greater understanding of the potential pitfalls, features, advantages and benefits of owning a piece of 21st century, in-use emission free technology.

Will petrol and diesel powered vehicles be completely replaced by electric vehicles? Many governments, including the UK, have pledged to ban the sale of all new petrol and diesel vans and cars by 2040. The main factor behind this date is almost wholly down to reducing harmful emissions to further improve the environment, and through reading this guide, I hope to highlight the impact that EV ownership will have on the future health of our planet, as we witness the transition to a carbon free world.

FACT

HEALTH BENEFITS. The expansion of EV's and general reduction of harmful emissions will lead to improved health and less noise on our roads.

1. THE HISTORY OF THE EV

The electric Vehicle is nothing new, they have been around for almost 200 years, long before the development of the internal combustion engine. EV's grew in popularity to a remarkable level at the beginning of the 20th Century, confronted extinction between 1910 and 1920, suffered almost complete oblivion following the arrival of the cheaper, mass market Model T Ford, then witnessed a renaissance around the 1970's, making a mainstream comeback in the 21st Century [1]. This is a snapshot into the often, neglected history of the EV.

The pioneering years

The first practical electric car came in the 1830's and was produced by the pioneering British inventor, Robert Anderson. His car ran on non-chargeable batteries and was smitten with many set-backs including maximum speed of 1.5mph and range of about two miles [2]. Then Holland, Britain and the United States produced some of the first EV's, but it wasn't until the latter half of the 1800's that British and French inventors created some of the first practical EV's, with greater range and speed.

At the turn of the 20th Century, as the height of the Industrial revolution took its grip and the west became more prosperous, there was more choice of transport alternatives in the newly developed motor vehicle market – now offered in electric, steam or petrol/gasoline derivatives. Thus, began a format war, with the internal combustion engine coming out the winner due to simpler, lighter and cheaper drive systems, the abundance of cheap fuel (gas/petrol) and the mass production of cars, pioneered by Henry Ford in the USA.

Whilst gas/petrol vehicles grew in popularity, the heavy impractical steam vehicles disappeared, and electric cars became a dying niche sector. EV's were practically silent, had no gears, so were simple to drive and didn't discharge noxious contaminants like steam or gas-powered vehicles at the time. Electric cars quickly became popular with city residents - especially women. By now, batteries were rechargeable and as access to electricity in the 1910's became more common, charging electric cars became more popular.

EV popularity sparked huge interest amongst many inventors and innovators. In the USA, Thomas Edison passionately believed that EV's were the superior technology and worked to build a better electric vehicle battery.

During the first 20 years of the 20th century, cities such as Amsterdam and New York each boasted a fleet of battery powered taxis. In the USA, almost a third of vehicles were electric cars. But cheaper energy in the form of petrol gas was starting to have an impact on the more expensive and less practical fledgling EV industry. One of the biggest setbacks that started the decline of EV's was that electricity was only available in cities, making longer distance travel, almost impossible.

EV Resurgence

Oil prices and petrol gasoline shortages reached a new level in the 1970's - climaxing with the Arab Oil Embargo in 1973, creating a mounting worldwide awareness in lowering the worlds dependency on oil [3]. Some Car makers began exploring opportunities in alternative fuel to power our cars, including the resurgence of electric car development again. Nonetheless, these newly developed cars or vehicles still suffered from the weaknesses of EV's in the early 1900's versus petrol gas powered cars. Range practicality and performance still rendered them undesirable.

The absence of public interest didn't dishearten the innovators, engineers and scientists from development in this revitalised new sector. Over the following 20 years, car companies tended to transform existing models into electric versions, in the hope that they could improve battery technology, achieving speed, practicality and range closer to their petrol gas powered relatives.

In the eighties, observers were closely watching developments in the UK as British inventor John Goodenough and his team at Oxford University developed the modern lithium-ion battery. It proved to be lightweight and energy dense. Exactly what the fledgling EV pioneers had been waiting for, in an effort to compete head on with the traditional polluting, noisy combustion powered vehicles.

2010 – 2020. A decade of intense innovation

Fast forward to the 2010's. One of the most significant turning points was the introduction of the Nissan Leaf, released in Japan and the USA in 2010 [4]. The Leaf became the world's first mass-produced pure electric vehicle. It slowly gained popularity held back only by the lack of charging stations. Since then, rising petrol prices and mounting concern about carbon pollution propelled the Nissan Leaf to become the best-selling pure electric vehicle worldwide during the last decade.

At the same time, new battery technologies evolved, helping to improve pure EV's range and lowering lithium-ion battery costs by more than 50 percent during the first four years. The battery had suddenly become a commodity and improved manufacturing techniques and new chemistry developments helped lower the overall end-cost of electric vehicles, making them more desirable and affordable for the car buying public.

Then came Tesla. A company, founded by Americans, Martin Eberhard and Marc Tarpenning [5]. They named the company after the Serbian inventor and developer of Alternating Current, Nikola Tesla. Although their first car, the Tesla Roadster was developed and built in the UK, based on an existing Lotus designed and built car, their breakthrough came in 2012, when they launched their first home-built Tesla S luxury saloon. Many commentators believe that the key to Tesla's success with the Model S, was the development and roll-out of Tesla supercharging stations, located in strategic locations across the USA on most of the country's main highways. A strategy replicated in all new Tesla markets to much acclaim.

Current market

As more EV's enter this new market, the charging infrastructure has become an important issue. Without a national contiguous network of fast or rapid EV chargers, the demand for EV's will remain muted. Although in most developed markets, the network is maturing and should maintain demand for many years to come.

In a major milestone for Estonia, it became the world's first country in 2009 to develop a nationwide EV rapid charging network.

The current state of global charging infrastructure indicates that Western Europe, the U.S, Japan and China, built more than 50,000 communal charging stations by 2014. Additionally, the EU enshrined in law, a directive to ensure a minimum coverage of charging infrastructure throughout EU countries [6].

By 2019, most major car manufacturers had adopted an electrification strategy. In the US and Europe, they have between 40 to 50 pure electric plug-in EV's. China tops the world ranking of most produced EV's, greatest network of charging points and the worlds largest fleets of electric taxis and buses [7]. They now lead the world in introducing EV's and electrified heavy commercial vehicles, reducing pollution in what is after all, the world's most polluted country, by quite a degree.

2. WHAT IS AN EV AND HOW DOES IT WORK?

This guide focuses on pure electric vehicles, also known as BEV's (Battery Electric Vehicles – figure 1.). This classification is often recognised as pure or fully electric vehicles, powered solely by an integrated battery pack that is charged from a slow charge domestic wall box, fast charge street charging post or rapid charger. For this reason, pure electric EV's are also referred to as plug-ins.

Generally, all EVs have either one or two electric motors rather than an internal combustion engine. Some EV's actually have combined motors built into their hubs, on all 4 wheels. All EV's use a large traction battery pack to power the electric motor and must be plugged in to a charging station or wall outlet to charge. Because it runs on electricity, the vehicle emits no exhaust from a tailpipe and does not contain the normal liquid fuel mechanics, such as a fuel line, fuel tank, or fuel pump.

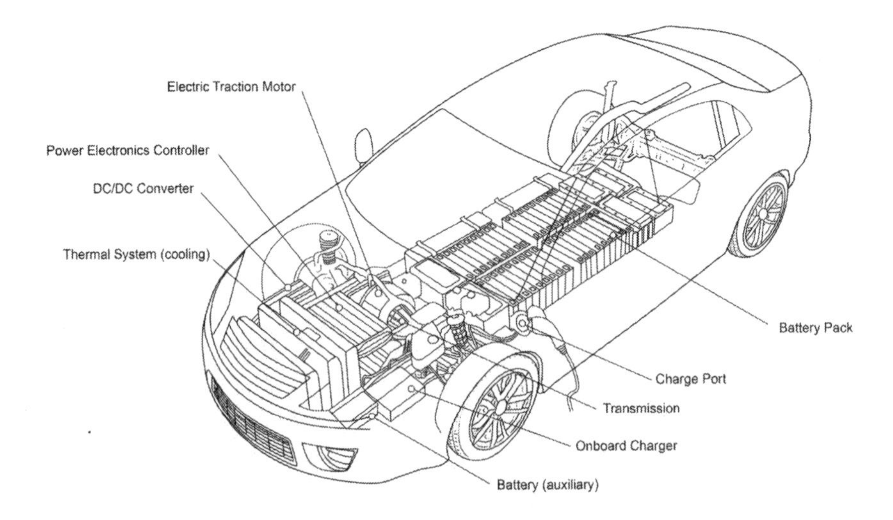

Figure 1 **Battery Electric Vehicle**

EV OVERVIEW

» **Auxiliary Battery:** In an electric drive vehicle, the auxiliary battery provides 12V electricity to power vehicle accessories.

» **Charge Port:** Enables the vehicle to connect to an external AC or DC power supply to charge the traction battery pack.

» **DC/DC converter:** This device converts higher-voltage DC power from the traction battery pack to reduce to lower-voltage 12V DC voltage needed to power the vehicle accessories and to recharge the auxiliary battery.

» **Electric traction motor:** Powered by the system battery pack, this motor drives the vehicle's wheels. Many new EV's are equipped with motor generators that function as both drive and regeneration/ braking.

» **Onboard charger:** Fed by incoming AC power supplied via slow and fast charge units, the onboard charger then converts AC to DC power for charging the main battery. It manages battery functions such as current, voltage, state of charge, and temperature whilst charging the main battery pack.

» **Power electronics controller:** This unit controls the flow of electrical energy delivered by the main battery, controlling speed of the main drive motor/s and the torque or power that it produces.

» **Thermal cooling system:** This system sustains the correct operating temperature range of the electric motor, power electronics, and other temperature sensitive components.

» **Traction battery pack:** Stores electricity for use by the propulsion system.

» **Transmission:** The transmission provides mechanical power taken from the electric traction motor to drive the wheels.

3. COMMON EV MYTHS AND FREQUENTLY ASKED QUESTIONS

A 2019 survey conducted by L.V. General insurance [8] in the UK found that one in ten UK motorists believe that electric cars cannot be driven in the rain!

This is just the tip of the mythical iceberg of misinformation, all unfounded and often inexplicable reasons preventing car owners considering the jump to a fully electric plug in battery powered vehicle.

Whilst full battery powered cars are increasing in popularity, LV have found that 27 per cent of drivers are considering switching to an electric vehicle within the next five years, both for cost benefits and for environmental reasons. Although an innumerable number of car drivers are still cautious about making the leap to an EV, due to mounting propaganda, coming mainly from the petrol and diesel lobby, who have a vested interest in preventing widespread adoption of zero emission electric vehicles.

To ensure car drivers that are thinking about buying an EV, are armed with unbiased facts, LV Insurance has dispelled some of the principal electric car myths below; some based on the responses in their 2019 survey.

MYTH 1 - Electric cars have no power compared to petrol or diesel cars.

According to LV, 55 per cent of UK adults believe an electric car isn't as powerful as petrol or diesel, although this is not true. Electric cars generally provide power much quicker and most accelerate far more rapidly than their petrol or diesel counterpart. In fact, the world's fastest road car is the Pure Electric Vehicle: Pininfarina Battista.

MYTH 2 - You can't drive an electric car on a highway or motorway.

An amazing 12 per cent of people are convinced that you can't drive an electric car on the motorway, although this is not true. There are more than 150 public stations at motorway and A-road services around the UK, providing more than 350 individual chargers. In the USA, this number increases to more than 5,000 DC rapid chargers.

On normal roads and towns, there are more than 20,000 public charging points across the UK, and in the USA the total to date is more than 50,000 and this number is expanding daily.

MYTH 3 - You can't use an electric car in a car wash.

18 per cent of people believe that you can't wash an EV in a car wash. Although electricity and water are not natural partners, it's completely safe to do so and all cars licensed for road use have to conform to stringent safety tests, the same as conventional cars do, including a soak test.

MYTH 4 - Electric cars and chargers shouldn't be used during heavy rain.

Not being able to drive an EV in the rain is a complete myth. All EVs have to be extensively tested by their manufacturers to prove compliance with a range of safety conditions, prior to being allowed on the roads.

Public and domestic EV chargers are weatherproof too, and all charge points must endure demanding safety testing. All charge points are deployed in strict accordance with the relevant electrical and safety regulations.

MYTH 5 - Electric car batteries need to be replaced every five years.

A common misunderstanding is that all EV batteries must be replaced five yearly, with more than a quarter of respondents believing this is true, according to LV's survey.

The truth is that current battery technology will last at least 10 years and feasibly even up to 20 years before needing replacement. Moreover, most manufacturers offer great warranties that prove beyond doubt, their confidence in the longevity of their batteries.

MYTH 6 - Electric cars are dangerous and unsafe.

LV's survey suggests that 6 per cent of respondents wouldn't buy an electric car because they consider they pose a danger and that there's a risk of electrocution. Of course, any electrical apparatus can be a potential hazard, but electric cars are at least as safe as a regular car.

Instead of flammable fuel like a standard vehicle, EV's are equipped with a lithium-ion battery power pack, just like a larger version found in a phone or laptop. These could ignite if misused, but EV manufacturers have installed devices to disengage the battery should there be a collision.

The agencies responsible for crash testing US and European vehicles respectively, National Highway Traffic Safety Administration and Euro NCAP, found that electric models like the Nissan Leaf, Tesla S and Jaguar iPace are exceptionally safe vehicles, awarding them all five-star ratings.

MYTH 7 - Electric cars can't be used for long journeys.

The LV survey also found that 45 per cent of people who said they wouldn't buy an EV, are discouraged from purchasing one because they think they are not capable of traveling long distances and consequently no good for long journeys.

Most of the latest breed of EVs now provide between 200 to 300 miles of range per charge in real world driving conditions – which is approximately the drive from New York to Washington DC or London to Manchester on one charge.

It is worth noting that you need to check the real-world range of electric cars, rather than manufacturers often inflated ranges, and the best guide is to use the latest world standard – WHLV test (World Harmonised Light Vehicle Test) rather than using the older NEDC (New European Driving Cycle) method of measurement.

MYTH 8 - Electric cars are more expensive to run.

Another myth according to the recent LV survey, is that electric cars are too expensive to run with 25 per cent saying they will not buy an EV because the perceived running costs are prohibitively high compared to a traditional petrol gas or diesel car.

But according to the Energy Savings Trust, with a full charge, an EV can run for 100 miles at a cost of around £4 to £6 or in the USA $5 to $7 USD, compared to 100 miles in a petrol gas or diesel car costing £13 to £20 in the UK or $10 to $18 in the USA (based on lower petrol/gas prices).

MYTH 9 - There are no incentives on offer for buying a new electric car.

LV found that a further 40 per cent of respondents in their 2019 survey were unaware that incentives were available when buying a new electric car. In the UK and mainland Europe, many governments have introduced a 'plug-in grant' – effectively discounting the price of brand new low-emission vehicles through a grant provided to vehicle dealerships and manufacturers. In the UK, the maximum current grant available for electric cars is £3,500. In the USA there are many state specific incentives available for BEV's, especially green states such as California.

MYTH 10 - Petrol and diesel cars won't be banned completely.

The LV survey found that 70 per cent of people were completely unaware that the sale of new petrol gas and diesel cars is due to be banned by 2040, and 15 per cent believed that electric cars will never entirely replace vehicles with combustion engines.

Yet, in a bid to tackle pollution, all new cars sold in the UK will be 'effectively zero emission' by 2040. Although there are moves for the UK Government to bring this deadline forward and ban all sales of new petrol and diesel cars by 2032.

MYTH 11 - EVs don't have much range and you will run out of battery power very quickly.

UK drivers average 20 miles per day, whilst Americans drive an average of 40 miles a day according to the UK government and US DOT [9]. Even the shortest-range electric vehicles currently available can travel more than twice that distance before needing to be charged by the mains supply or alternative power source. Amongst affordable EVs, the Nissan Leaf averages 150 miles per charge, while the Chevrolet Bolt EV is even better with a claimed 238 miles and the battery only version of the Hyundai Kona and Kia E Niro boasts a real-world range of 258 miles each. For the more affluent amongst us, the Jaguar iPace provides 250 miles on a charge, while the premium version of the Tesla Model 3 has 310-mile range. In the near future, many EV manufacturers are quoting 400mpc (Miles Per Charge) plus on their new models.

MYTH 12 - EVs are as slow as golf buggies or milk floats.

This is a truly unfounded myth, based on people's perceptions of golf buggies and milk floats. Electric vehicles are generally faster than their petrol gasoline-powered competitors. This is because an electric motor provides 100% of its available torque instantly. When the driver of an EV floors the accelerator pedal, the conversion from stationary to high speed is almost immediate. At the fast end of the production car scale, the top range version of the Tesla Model S, when switched to its "ludicrous" mode, is one of the fastest production cars in the world. It has a 0-60 mph time in the fastest mode of an incredible 2.5 seconds.

MYTH 13 - EV's are really expensive.

The single most expensive component in an EV is its battery pack [10]. Although costs are expected to drop dramatically over the coming years, at the moment, this one factor is slotting most EVs in the premium price bracket, compared to similar petrol gas-powered or diesel cars. But most EVs are eligible for a one-time grant. In the US there is a $7,500 USD federal tax credit granted to EV buyers, helping to introduce an equilibrium in comparative pricing. In the UK and many other European countries, there are 'plug in' grants available, which in the UK is currently set at £3,500.

A few US states offer specific grants to EV buyers. California residents can get a cash rebate of between $2,500 and $4,500 from the state, which is means tested. While in Colorado, residents are eligible for a $5,000 state income tax credit.

Of course, If you want to drive an EV but are on a tight budget you can consider a used model. Pre-owned EVs are fairly low cost. Also, used electric cars tend to be driven fewer miles than their gas-powered counterparts, given the inherently limited ranges in older models, with the added bonus that they've typically suffered less wear and tear.

MYTH 14 - EV's are no greener than petrol or diesel cars.

Internal combustion engined vehicles (ICEs) only convert 20 percent of the energy stored in petrol gas. By comparison, electric motors convert on average, 75 percent of the chemical energy from the batteries to power the wheels. Furthermore, EVs produce no direct tailpipe pollution. Some critics and frankly, sceptics, argue that EVs still pollute the atmosphere indirectly, via the power plants that produce the electricity necessary to operate them. But few of these critics take in to account the huge volume and waste of electrical energy that is used to refine petrol gas and diesel.

Recently, there was a whole week when the UK used absolutely no coal powered energy at all. Many countries are heading towards the goal of 100% renewable energy [11]. A study conducted by the Union of Concerned Scientists [12] cited that EVs are generally responsible for less pollution than conventional vehicles in every region of the U.S and this is generally the case across the UK.

MYTH 15 - Based on today's petrol gas and diesel prices, driving an EV will not save you money in operating costs.

Even if the cost of gas remains relatively affordable, especially in the USA, it's still much cheaper to run an EV. For example, in the USA, the Environmental Protection Agency (EPA) maintains that the Hyundai Ioniq Electric will cost its owner $500 a year to travel 15,000 miles, based on average electricity costs. This amounts to an estimated $5,000 less than the average conventional vehicle owner will spend in fuel costs over a five-year period. In the UK and mainland Europe this saving is at least doubled.

MYTH 16 - Electric cars are expensive to repair and maintain.

Compared to the increasing complexity of conventional cars and the subsequent high cost of maintenance, EVs cost less to keep running than ICE-powered vehicles as they don't require routine tuning or oil changes and there are far fewer moving parts to eventually fail that need replacing. Most EVs use a simple single-speed transmission and avoid consumables such as spark plugs, valves, fuel tanks, muffler/tailpipe/exhaust systems, distributors, starters, clutches, drive belts, hoses, and catalytic converters.

MYTH 17 - The EV charging infrastructure is not adequately established yet to make owing an EV an attractive proposition.

Generally, electric vehicle charging is done at home or at work. Although, if charge is needed during the day, you can usually find charge points at retail car parks, public parking garages, and new car EV dealerships. While most chargers in town are 230-volt type 2 fast chargers, that take around four hours to replenish an average EV's battery pack, a mounting number of DC Rapid Charging points are also coming on stream. Rapid chargers can replenish to 80% of an EV's state of charge in about 30 minutes (depending on battery pack size and state of charge of your vehicle). For long distance trips, its essential to map a course to your destination that's covered by Rapid chargers.

MYTH 18 - EV battery packs have a limited lifespan and will end up in landfill.

In the USA, EV's are federally required to carry separate warranties for their battery packs for at least eight years or 100,000 miles [13]. In the UK, most EV manufacturers offer battery warranties for between 8 to 10 years. Published reports suggest that Nissan Leaf models used as taxicabs, retained at least 75% of their battery capacity after 120,000 miles on the road.

Once exhausted, EV batteries, similar to 99% of the batteries found in conventional cars, can be recycled. For example, used EV power cells can be and are used to store solar and wind energy, or they can be recycled to reuse their more-valuable elements, such as lithium.

MYTH 19 - The power infrastructure and grid can't cope with the forecasted additional EV's on the road.

According to a report conducted by Navigant Research [14], the UK can add millions of electric cars to the current power system with little impact on the grid and without having to construct any new power plants. The main reason for this is that most electric vehicles tend to be recharged at night during off-peak hours when the grids power demand is at its lowest point. There have been similar reports across Europe and the USA.

FACT

Electric cars are better for the environment, regardless of the electricity source used.

FACT

The world's fastest road car is pure electric reaching a speed of 217mph (350kph). It is the Pininfarina Battista.

4. EV TECHNOLOGY

EV VARIANTS

There are five main variants of vehicle that are loosely grouped under the umbrella of the term EV or Electric Vehicle. In its purest form, there is the BEV or Battery Electric Vehicle that is powered solely on recharged batteries. These vehicles are rapidly becoming the first and greenest choice of electric vehicle. The following EV types are listed chronologically in their green ranking, with most efficient first.

FACT

AVAS. In the EU, from July 1st 2019, all 4 wheel electric cars must be fitted with an AVAS (Acoustic Vehicle Alert System) at low speed, following concerns that pedestrians were at risk because the cars could not be heard as they approach.

BEV - Battery Electric Vehicle

A BEV is powered by either 1 or 2 indirect electric motors, or by 4 integrated hub motors, each controlled by an electronic motor controller that manages functions such as limited slip differential, torque, regenerative braking and incremental speed shifts.

A BEV obtains its electric power by plugging in to a public or home charger that is normally connected directly to the electrical grid or via a BESS (Battery Energy Storage System) often used to collect energy from Solar, wind or other alternative energy sources. This variant of EV uses no direct fossil fuel to power it, nor is there any tailpipe emissions. Typical examples include the latest Nissan® Leaf, BMW® i3 BEV, Chevrolet® Volt, Tesla® S and Jaguar® iPace.

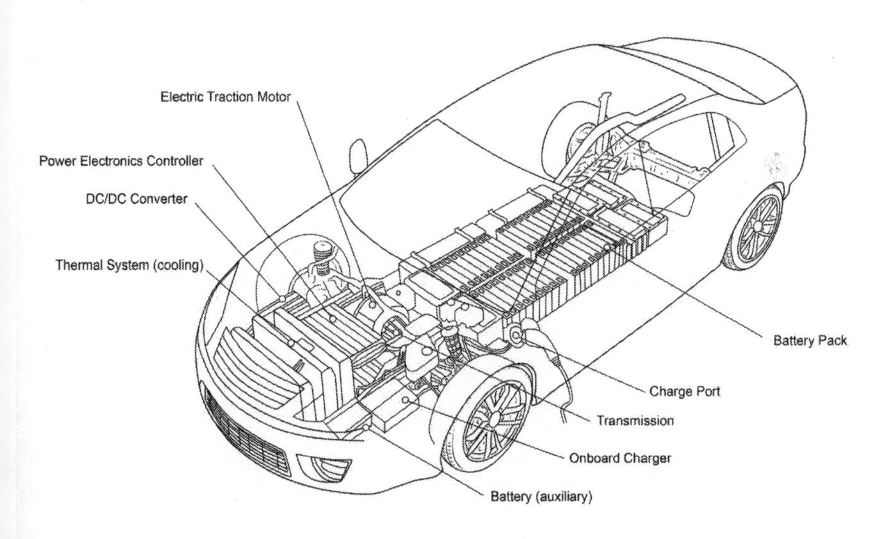

Figure 2 **Battery Electric Vehicle**

FCEV - Fuel Cell Electric Vehicle (Hydrogen)

The FCEV type of EV is powered by a hydrogen fuel cell. This normally delivers a steady current to the cars in-built battery pack, through a battery management system and then the car can be driven in the same way that a pure battery powered car is operated. The FCEV has similar green characteristics to the BEV as it also emits no tailpipe pollution, just clean water – the by-product of the chemical reaction caused by the fuel cell converting hydrogen and oxygen from the air into electricity.
On some FCEV's the electricity is then stored in onboard batteries similar to the BEV and the stored power is used to regulate the power driving the motors of the EV.

The hydrogen is stored in secure tanks within the car and it needs to be refilled at one of just a handful of nationwide refilling stations. The principle of Fuel cell powered EV's is similar to REHEV cars, since the fuel cell often acts a generator of electricity to charge the small stack of batteries. On other FCEV designs, the fuel cell drives the motors via a motor control system direct.

Nevertheless, there can be a modicum of noise pollution with some fuel cell powered vehicles, due to the noise of a compressor used in the design, depending on the variant of vehicle. Generally, FCEV's are more complex than simple BEV's and demand much more maintenance. Furthermore, charging station infrastructure is scarce at the moment. But the main benefit is that there is no tailpipe pollution with either variant of FCEV. Typical examples of FCEV include Hyundai® NEXO, Toyota® Mirai and Honda® Clarity.

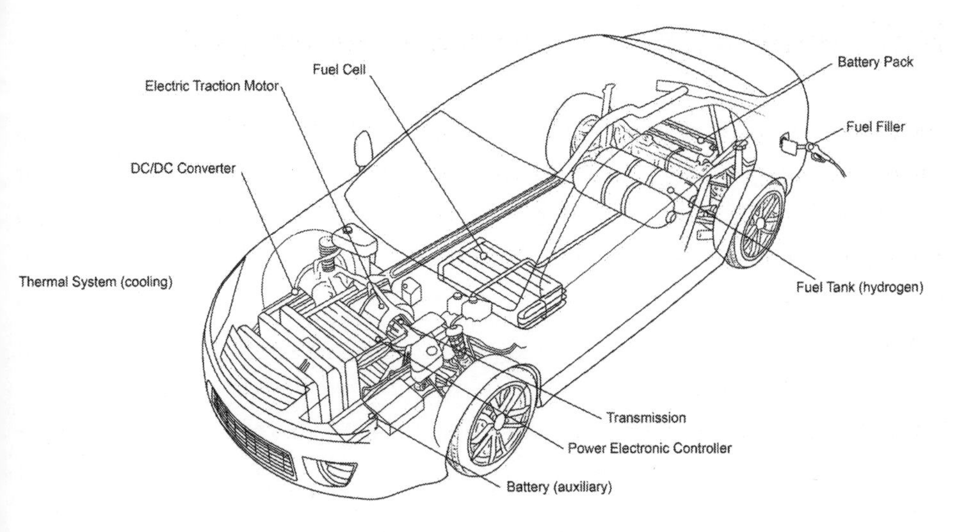

Figure 3 **Fuel Cell Electric Vehicle**

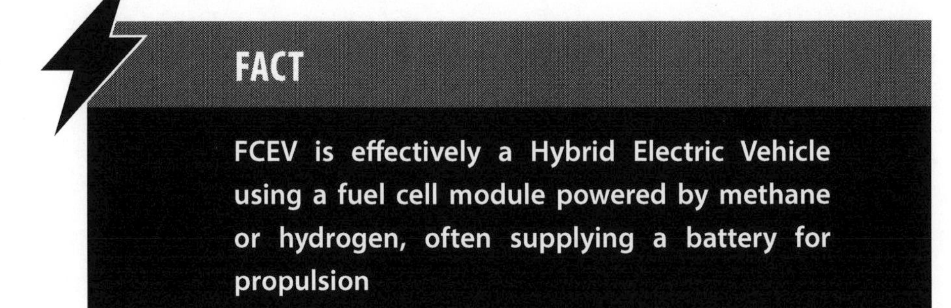

PHEV - Plug in Hybrid Electric Vehicle

PHEV's employ a small lower voltage battery pack compared to BEV's and are used to power an auxiliary electric motor that is linked to the conventional transmission system or in some cases, powers the rear wheels, whilst the main ICE (Internal Combustion Engine) powers the front wheels. The in-built battery storage is much smaller than a BEV and provides much lower 'all electric' range compared to a BEV.

A PHEV obtains its electric power by plugging into a public or home charger that is normally connected directly to the electrical grid or via a BESS (Battery Energy Storage System) often used to collect energy from Solar, wind or other alternative energy sources.

The PHEV's main energy source is from Petrol Gasoline or Diesel, to power its internal combustion engine. This EV variant does produce significant tailpipe emissions and can normally only be driven on battery power for short distances ranging from 6 to 30 miles, depending on the model. Typical examples include Mitsubishi® PHEV, Range Rover® Sport PHEV, Audi® A3 PHEV and Cadillac® CT6 PHEV.

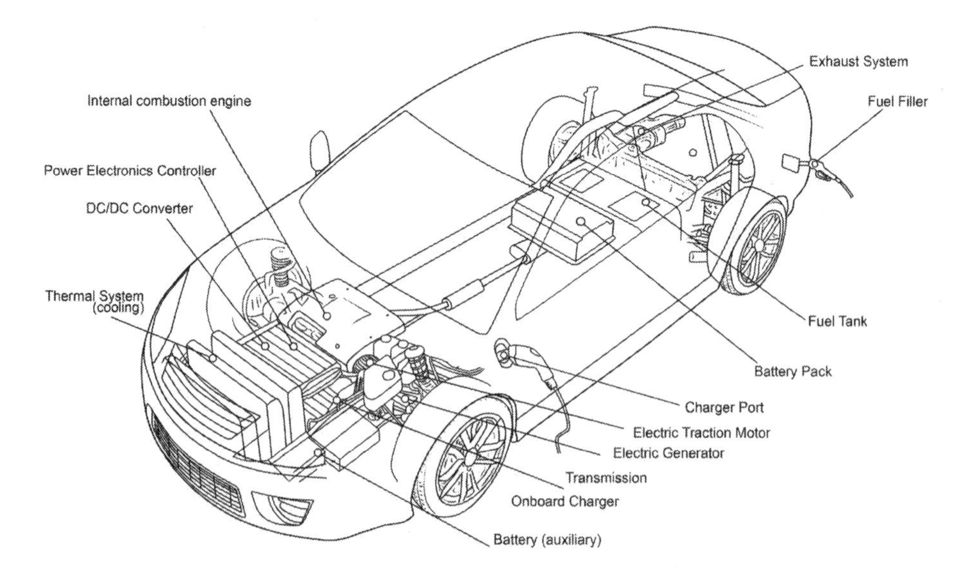

Figure 4 **Plug in Hybrid Electric Vehicle**

HEV - Hybrid Electric Vehicle

The HEV variants combine battery storage with an internal combustion engine. This EV type has an electric auxiliary motor, normally assisting the main engine via a link to its main transmission system. It often comes with regenerative braking.

HEV's rely on petrol gasoline or diesel for their main source of power and have no external facility or function to enable its battery pack to be charged by an external charger. The batteries are most commonly charged by the engine and also during the braking process. This EV variant produces significant tailpipe emissions and can normally only be driven on battery power for short distances ranging from 5 to 20 miles, depending on the model. Common examples of this car type are all Toyota® hybrids, Lexus® hybrids and Hyundai® hybrids.

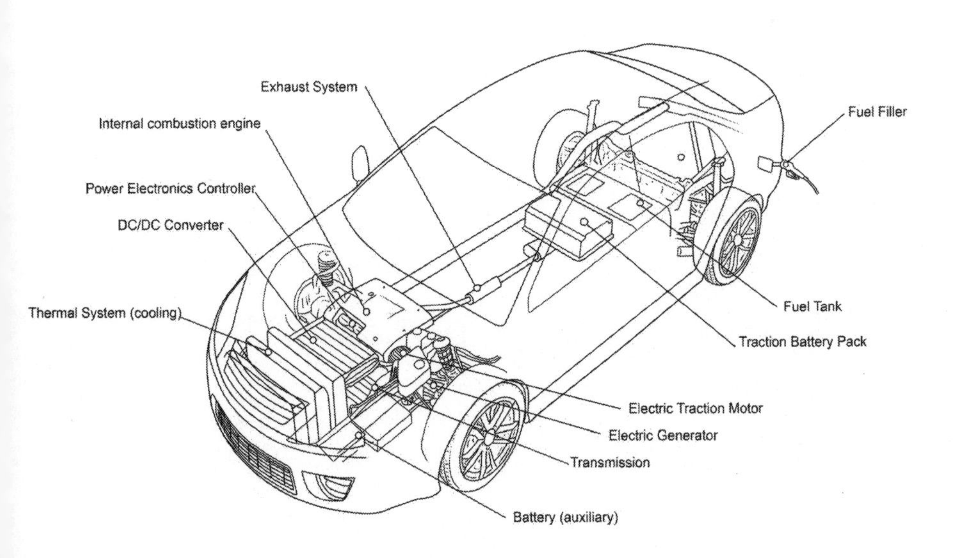

Figure 5 **Hybrid Electric Vehicle**

REHEV - Range Extender Hybrid Electric Vehicle

REHEV variants are similar to a pure BEV with the added range security of a petrol gasoline powered generator. The generator is not connected directly to the drive train, instead it charges the battery pack direct.

This EV variant is losing popularity now and when used in range extender mode, produces tailpipe emissions. Examples are BMW® i3 REX, Vauxhall® Ampera REX and Chevrolet® Volt REX.

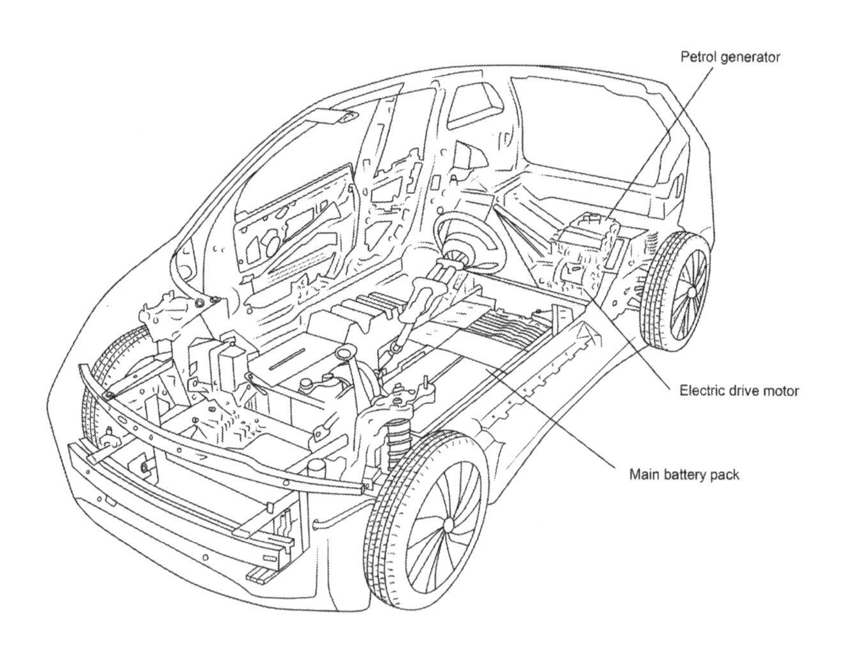

Figure 6 **Range Extender Hybrid Electric Vehicle**

BATTERY TECHNOLOGY

There are two types of battery: primary and secondary. Primary batteries have a limited lifespan producing pre-stored energy and can generally not be recharged. Secondary batteries installed in electric vehicles are rechargeable cells that store electricity once recharged. Batteries are normally positioned down the centre of the chassis with more mass at the rear, to provide better safety and weight distribution. Batteries for electric cars were lead-acid (LA) for over a century, until Nickel Metal Hydride batteries (NiMH) gained favour for their lower weight and high current or charge density. But now, almost all modern electric cars use some variant of lithium battery (Li-ion).

The mass and weight of electric cars has always been a frequent design issue. The battery and electric propulsion system typically account for 40% of the weight of an EV, while in a fossil fuelled car, the engine, coolant system, and ancillary devices only amount to 25% to 30% of the weight of the car, dependant on fuel tank size.

Newer technologies under development, might however deliver alternatives that are more acceptable to the general public and like BEV's, may be zero emission at use too. For instance, fuel cells EV's have a chemical source of either hydrogen, ethanol, methane or even petroleum gasoline that produces electrons that generate electricity to drive the power train. A FCEV's down side is that there is no contiguous national hydrogen filling infrastructure yet in any country and then of course there is the safety aspect. But because the vast majority of fuel cells produce a constant voltage, then they need a pack of batteries to soak up the fuel cells output and store for a more controllable output to power the EV. Lithium batteries are now the 'de-facto' battery of EV choice, but their success may still be hampered by limited supplies of raw materials needed to cope with the huge demand on battery manufacturers as the world heads for a green, carbon free future.

Lithium batteries

There are 5 main lithium technologies in common use on EV's today, with two of these used in most electric cars, Lithium Nickel Manganese Cobalt (NMC) and Lithium Iron Phosphate (LFP). The following section focuses solely on lithium technology used in EV's and highlights how a simple lithium cell charges and discharges, summarising the main lithium chemistries, strengths, weaknesses and features.

How a lithium battery works

The diagram below (figure 7) illustrates the basic function of how a lithium battery works in its most basic form. The graphic is expanded to clearly show the cells constituent parts. Although, in reality this cell can take on a two main forms:

» **Cylindrical:** similar in style and look to a normal AA flashlight battery

» **Prismatic:** this is where the 5 different layers are sandwiched and protected in a thin ribbon, protected normally by an alloy pouch.

FACT

EV car batteries can be recycled and used for many more years in BESS (Battery Energy Storage System) to store energy at off peak times to reinforce the grid.

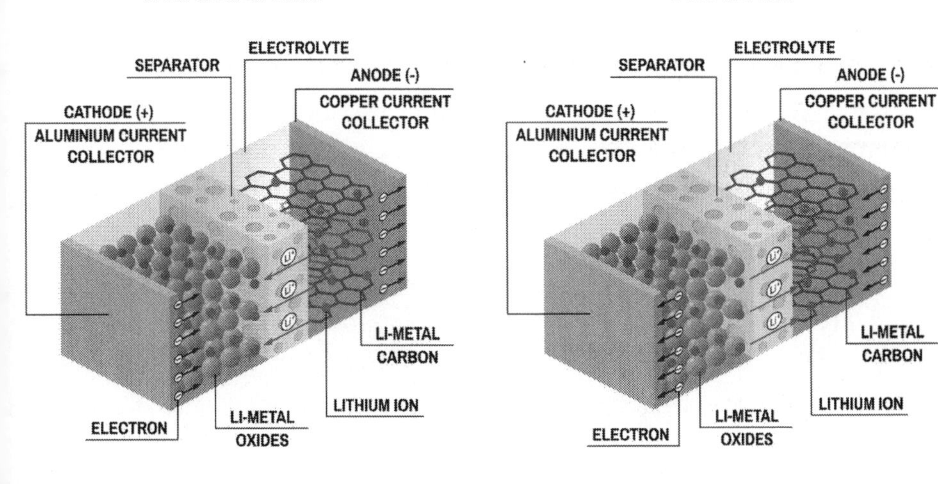

Figure 7 **How a Lithium Battery Works**

Discharge. When a lithium cell is discharging, the lithium-ions move from the negative electrode (anode) across the electrolyte (Viscous chemical mixture) to the positive electrode (Cathode), producing the energy that powers the battery.

Charge. When the battery is charging the chemical mix of metal oxides, the positive electrode sends over some of its lithium-ions, which flow through the electrolyte to the negative electrode (anode) and remain there. The battery consumes and stores energy in this way, during this charging process.

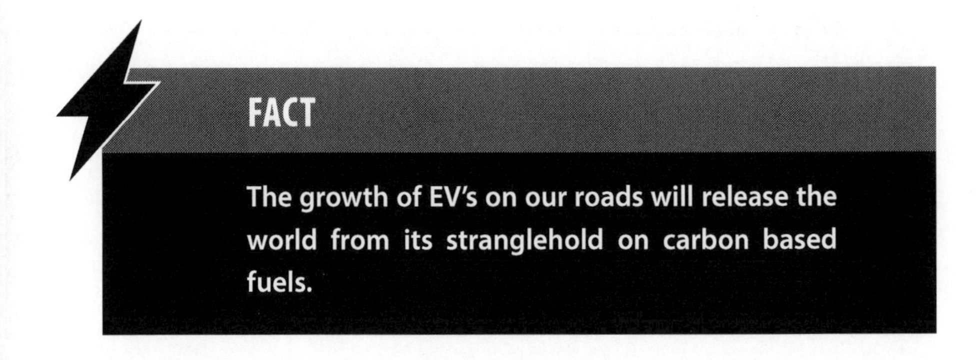

Cost, safety and energy density

From a manufacturing perspective, there are four key areas of uncertainty. All have a direct impact on price, safety and cost to produce and cost to market. One thing is absolutely certain, that battery costs will fall over the coming years with technological advances, increased production, modular and charging standardisation.

What influence will the cost of developing differing types of batteries, have on the developing EV market? What will be the scale of progress over the coming decade and what are some of the key obstacles that will need to be addressed?

» Cell chemistry development – evolving technology
» Rare earth material availability – key components in lithium battery technology
» Uncertain scale of consumer and commercial demand
» Policy direction from national and local governments
» Without government aid, there is a limited amount of research and development capacity. This can affect long term reduction in cost, higher density capacity and overall cost of EV's in general.

To highlight these and many other issues that all manufacturers of EV batteries must address as a matter of course, the following diagram (figure 8) reveals ten critical battery characteristics that allow positive battery design, innovation, safety, performance and environmental requirements to be met exponentially.

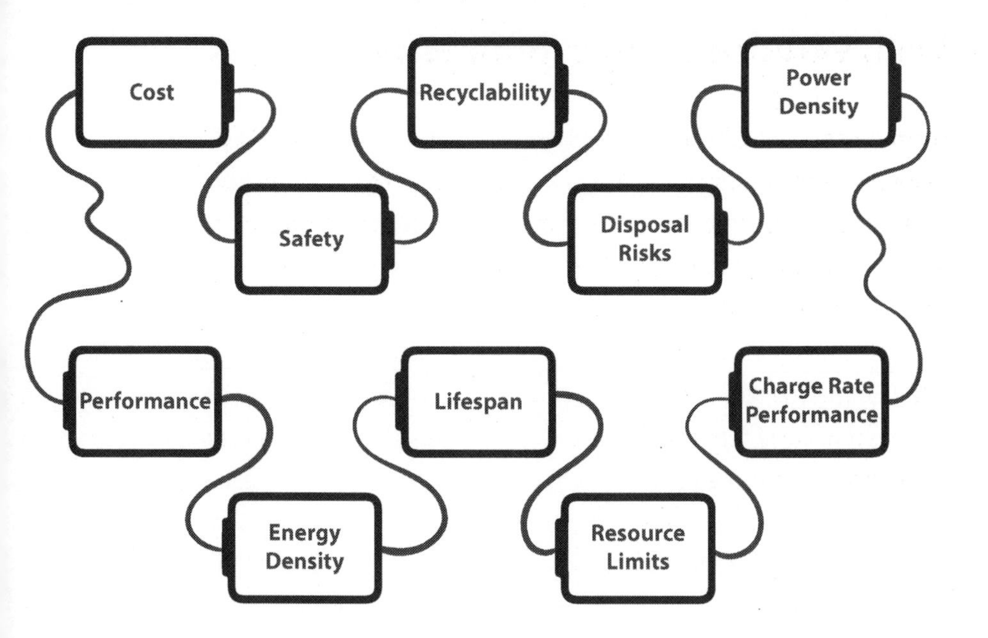

Figure 8 **Battery Characteristics**

FACT

Rapid battery recharging is quick and easy. There are more than 5000 rapid chargers in the USA and almost 2000 in the UK.

Main variants of lithium battery technology

The recent boom in lithium-ion based battery innovation is powered by the industry thirst to conquer major battery technology compromises. There are six aspects in the technical areas of battery design that are competing with one another (figure 9). Ideally, all 6 aspects of each battery chemistry would score high (outer edge of each chart) in all six areas. The closer to the centre of each chart a parameter is set, the lower its performance is in that area.

» **Performance:** Peak power at high and low temperature, thermal management and state of charge analysis.

» **Lifespan:** Measured in how many years the battery will operate within the design parameters and the number of charge cycles.

» **Specific power:** The amount of power the battery can store per kg of weight. The higher the safe speed of energy discharge, the higher the power rating.

» **Specific energy:** The amount of energy the battery can store per kg of weight.

» **Cost:** The challenge over the coming years is to reduce battery manufacturing costs, by R&D progress and scale of manufacturing to produce higher volumes. Economies of scale are key in this area as the market grows.

» **Safety:** The safety characteristics built into the design of the chemistry used in each technology. Currently LFP and NCA offer the highest degree of safety, although solid state batteries promise even higher levels of safety in the near future. There are design countermeasures built-in such as intumescent boxes or cell pack coatings to prevent the spread of heat or flame.

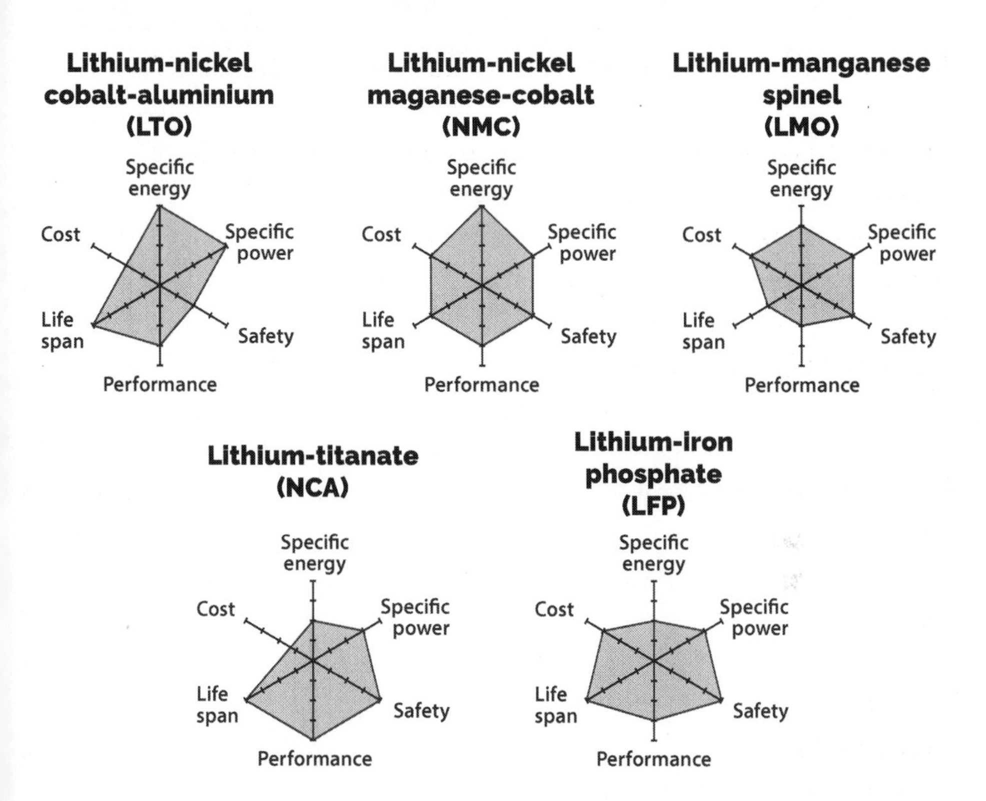

Figure 9 **Lithium Battery Variants and Trade-offs. Source: BCG [15]**

Figure 9 [15] reveals that there is no individual technology that currently achieves top marks in all six key battery factors. Therefore, the choice of battery chemistry and technology by each EV manufacturer is a compromise on one or two aspects in an effort to optimise the operational function of an EV, versus overall cost and EV performance. This issue is no different to a conventional combustion engine, where although the engine principal is the same, all combustion engines perform differently, look different and have similar overall performance pros and cons as EVs.

Range

EV range depends on the quantity and variant of batteries used, the type and weight of vehicle, weather, road terrain and performance requirement. The real-world range of production electric vehicles in 2019 ranged from 62 miles (100km) to 340 miles (540km) specifically the Tesla Model S 100D [16].

All electric cars are fitted with a display that forecasts range, normally taking into account the way the vehicle is being driven, and what ancillary electrics (such as Air conditioning or heated seats) the battery is operating. Nevertheless, since influences on battery drain can vary on route, the estimate can differ from the actual range. The display simply allows the driver to make educated choices about driving speed and whether or where to stop at a charging point on the way.

Lifespan

All lithium-ion batteries used in EVs will degrade slowly over time, particularly if they are frequently overcharged, though, this may take several years before it would be noticeable.

In 2015, after 10 years of data and feedback from dealers and drivers, Nissan indicated that only 0.01 percent of batteries needed to be replaced through failures or problems and in all cases, only due to externally imposed damage [17]. Both Tesla and Nissan have vehicles that had already travelled in excess of 124,274 miles (200,000 km), with no significant battery issues. In fact, the vast majority of manufacturers now guarantee their battery packs for a minimum of between 80,000-100,000 miles (130,000km - 160,000km) or 8 to 10 years, such is the confidence in their battery life.

Future battery innovation

Innovation in lithium-ion battery technology continues to develop at a fast pace. Innovations include material development for cathodes and anodes, new cell coatings and protection, electrolyte development, processing of electrodes and new design and innovation in the assembly of batteries. Other developments include innovative liquid and solid electrolytes, leading to new solid-state batteries and graphene technology including future super capacitors, forecast to replace batteries in the future.

Solid state batteries

Solid state battery technology is viewed as the Holy Grail of stored power. It has been under development by research institutes and battery manufactures for more than a decade. There has not been a major breakthrough yet in achieving economically viable performance on a commercial scale.

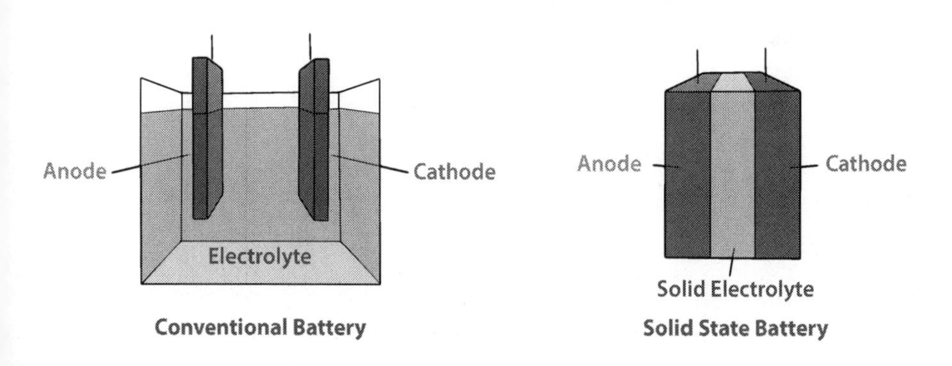

Figure 10 Lithium-Ion Versus Solid State Lithium

Solid state batteries use much lighter polymers to store and discharge energy. This technology promises far lighter and less costly power storage that potentially negates the use of rare earth metals and environmentally damaging manufacturing processes. In the past, solid state battery prototypes had not demonstrated the ability to be scaled-up commercially to employ in

EV's [18]. This obstacle has meant that solid state battery research has seen huge investment in time and capital to ensure that these new batteries offer significantly advanced performance over traditional lithium-ion batteries. But the most important criteria, is the target to create a solid state battery, that is economically viable to manufacture and practical enough to be scaled up for use in EV's, to survive the future move towards carbon free transport.

Super capacitors

Super Capacitors, or Ultra Capacitors as they are also known, potentially have significant advantages over batteries. They are faster to charge, much lighter, non-toxic and safer. But at the moment, they are not quite ready to compete head on with traditional lithium-ion battery types [19].

Like batteries, super capacitors are a type of energy storage device and they have been gaining a growing reputation over the past few years. In effect, the super capacitor is a hybrid capacitor/battery, but in reality, quite different from both. Unlike batteries, super capacitors store energy like a conventional capacitor, electrostatically, rather than chemically like a battery.

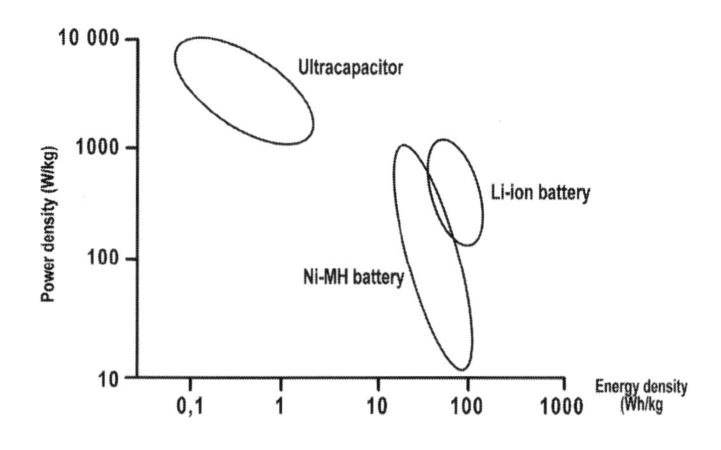

Figure 11 **Comparison of Ultracapacitor and Battery**

Nonetheless, compared to batteries, super capacitors store less energy than a similar mass battery in their present guise. Though they can release energy much faster than a battery, since the discharge is not reliant on a chemical reaction happening. The other great benefit is that super capacitors can be charged many more times than batteries with little or no degradation. In fact, under controlled testing, they have been known to exceed more than one million charge and discharge cycles, mainly because no chemical or physical reaction is occurring during a charge or discharge cycle. Once perfected, this technology holds great promise for the future of EV's.

Graphene technology

Since graphene was first discovered in a laboratory at the UK's Manchester University in 2004, developments have moved fast towards applications to replace steel (it is 10 times stronger than steel but 1000 times lighter than a sheet of paper of the same dimensions) [20]. But by far one of the most promising developments has been its potential for ultra-high-density batteries for EV's. The main advantage of using graphene for a battery electrode is that it has a high surface area and yet can be just one atom thick. To make this concept practical, the layers of graphene must be packaged into a 3D object. This means that in future, car bodies and doors, could also be the cars batteries. This technology is still in its infancy. It is not expected to reach market for at least a decade.

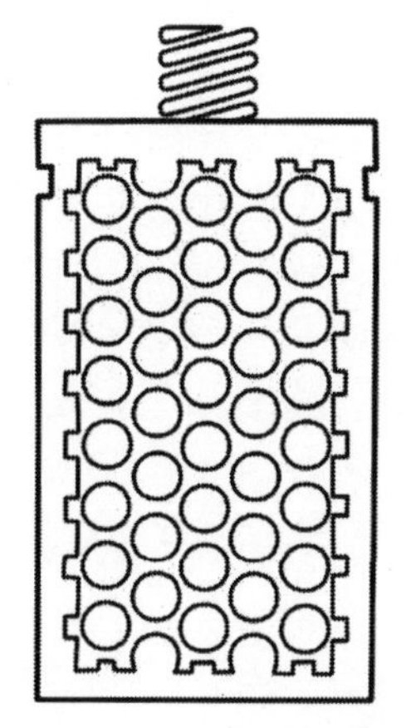

- **Higher temperature range**
- **Faster charging**
- **Higher capacity**
- **Lightweight**
- **Flexible form factor**

Figure 12 **Graphene Battery Advantages Over Lithium-Ion**

ELECTRIC MOTORS: DRIVE TRAINS OF THE FUTURE

Generally, the first difference you will notice between EV's and ICE vehicles, is that there are no gears to operate as such and are in effect automatics, with the vast majority of EV's designed as 'direct drive' rather than powered through manually geared transmissions.

Rapidly evolving EV motor technology is redefining and setting the electric vehicle landscape. EV motors account for less than 10% of the total car cost. There are many new and established companies entering the EV propulsion system market. One such new EV entrant is the Dyson corporation, for years dominant in the vacuum cleaner, hairdryer, fan and hand dryer business, centred primarily around its expertise in its motors. As part of a 2.5bn global investment, Dyson has now set up its first electric car manufacturing plant in Singapore [21].

Aside from Tesla, one of the main reasons that no other company has scaled up its EV manufacturing output, is that there is an incapacity to match the powertrain efficiency on a small scale. Tesla's key to its success in part has been its development and use of a modified version of the switched reluctance motor. This new breed of motor delivers very high power density at low cost, making them very appealing for EV propulsion and transmission systems. Though, there are drawbacks too, including high torque ripple: in one revolution, the difference between maximum and minimum torque often creates difficulties in precision low speed control and may cause unwarranted noise, especially when operated at low speed.

There are some commentators who forecast that precision EV component costs will decrease fast over the next decade to the point where commoditisation allows them to be obtained from low skill bases such as China. But most industry experts believe that this will not happen, just as jet engines remain in high skill manufacturing bases after almost 100 years. This may point to batteries becoming a commodity as production techniques and costs fall, but motor development and R&D costs may well level out as more efficient power trains become the norm.

There is an emerging trend of propulsion systems available in the EV market now. For example, some new four wheel drive EV's use an electronic differential and place motors with integrated hubs on all four wheels, lowering efficiency losses and simplifying the power train design. There is also a growing trend of integration in the form of 'motor in axle' solutions. Both of these solutions reduce overall cost, increase efficiency and lower complexity of the current generation of EV's.

Currently there appears to be two distinct choices in motor technology: the switched reluctance motor option, and advanced permanent magnet motors, now specified in many e-axles solutions. The more immediate breakthroughs can be seen in the rising use of Permanent Magnet motors, presently specified for more EV's than reluctance motors [22].

There is a growing trend to move away from one to four traction motors in all electric vehicle types. Certainly, growth in motor types, efficiency and integration is powering a myriad of options for the consumer, seemingly making your choice of EV an impossible task. Nevertheless, you can be assured that with few exceptions, car manufacturers and their designers will provide you with the best possible integrated system for your new vehicle

Regenerative Braking

Most new BEV's use a clever technology known as regenerative braking. This technology uses the momentum and mass of the EV to charge the battery pack when you brake. Most regenerative braking also enables the driver to literally drive with a one pedal operation. When you take your foot off the accelerator, the regenerative braking initiates, slowing the car down as the vehicle reverts to power generation, charging the battery and creating electrical drag on the drive system, providing you with zero cost power and physical disc free braking whenever you release pressure of the accelerator.

Almost all new generation BEV's integrate a recuperation function that enables you to adjust the driving profile of your car. This facility enables you to regulate the intensity of your braking and its harnessing of power through regeneration. In essence, when set correctly, it is feasible to drive with just the accelerator pedal, known as 'one pedal driving'. See figure 13.

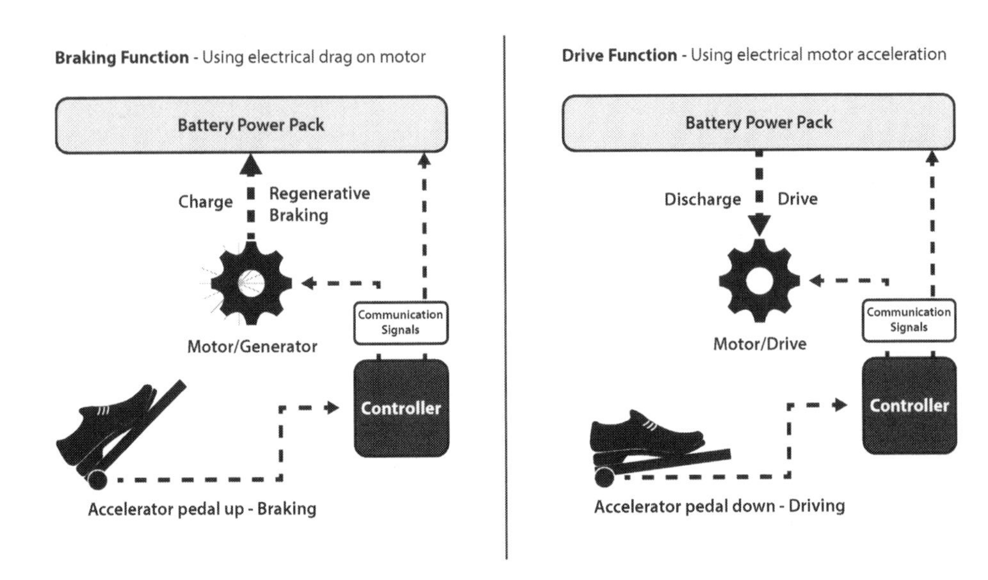

Figure 13 **Regenerative Braking - One Pedal Driving**

Drag Coefficient and Aerodynamics – does it really matter?

In reality, only cars designed primarily for road use really need to reduce drag force, particularly in this era of low carbon output and overall fuel efficiency. The main factor that determines whether a car is aerodynamically efficient is known as the drag coefficient, or CD. This calculated number simply provides a value to how effectively a vehicle can force its way through air. Currently, the best performing cars are averaging 0.26 CD, while most generally sit between 0.30 and 0.35.

Aerodynamic drag increases with speed in any object; it therefore becomes key at higher speeds. The higher the speed, the more aerodynamic drag. By decreasing the drag coefficient in a car, the designer improves the performance of the vehicle as it relates directly to speed and fuel efficiency. There are numerous methods to reduce the drag of a vehicle within the design phase of development.

So, should we be heading towards the magic, if not impossible 0.1 CD number? Well not exactly. A degree of drag force is very much needed to maintain grip on all four wheels in a controllable manner. Additionally, we need to ensure that some airflow is forced on to the heat exchangers that your car may have to maximise cooling of the battery packs, for example. It appears that the conventional ICE motor industry has taken the pragmatic view, producing cars between 0.30 and 0.35, from a design and practical viewpoint. But electric vehicles need to extract every last Volt of electricity to increase its range. Thus, any CD factor below 0.30 will reward the owner by significantly reducing your monthly charging costs, and as a by-product of better CD design, will allow increases in potential top speed.

Body design and construction

Car manufacturers are masters of the phrase 'if it isn't broke, don't fix it'. Most simply enhance their car frame designs over years of research and development to prevent a complete redesign for a new EV and this throws up all sorts of compromises. Conventional design has been optimised for Combustion engines, exhaust systems and geared transmission systems.

Though, EV's require a completely new way of design approach to the benefit of end-users. The by-product of custom-made designs for EV's is that you will get more interior room and expect to see EV costs continue to fall over the coming years as OEMs develop better, more efficient means of producing them, with lower battery prices starting to make an impact on forecourt prices.

5. CHARGING INFRASTRUCTURE

CHARGING PROTOCOL AND PLUG HARMONISATION

Currently, car makers in Europe, China, Japan and Tesla in the USA, use two communication protocols and four different connectors to link chargers to batteries, but companies constructing the charging networks necessary for electric vehicles to become mainstream, say the number of plug formats will need to be limited and even consolidated to one main protocol to keep future costs low.

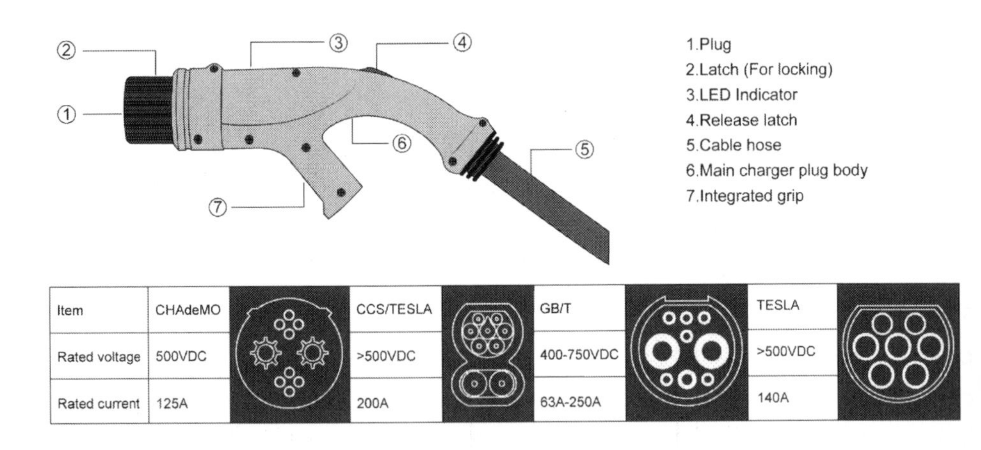

Item	CHAdeMO	CCS/TESLA	GB/T	TESLA
Rated voltage	500VDC	>500VDC	400-750VDC	>500VDC
Rated current	125A	200A	63A-250A	140A

Figure 14 **Rapid Charger Connector Variants**

It is widely thought that over time CCS and CHAdeMO will converge, most likely into the current CCS protocol, Tesla looks to be moving this direction too, with all European Model 3's fitted with CCS connectors. The main benefit of CCS over its competition is that it is a dual mode plug, capable of using Type 2 AC charge connectors, as well as the CCS combined AC and DC plug. Figure 1⁴ the stark differences between the 4 global rapid charge formats.

At the moment, Tesla Supercharger, CHAdeMO and CCS plugs continue to be installed in Europe as well as the United States and China appears to be powering ahead with GB/T, implying that it is too early to declare the victor in this war of plugs.

CHARGING STATIONS

Pure battery EV's (BEV's) are most commonly charged from the electricity grid overnight at the owner's residence, so long as they have their own charging box or adaptor to plug directly into the mains sockets. Power from the grid is generated from a mixture of sources. The main generators use nuclear power (arguably green power), coal, and gas. Although renewable power is becoming more common in most countries, including hydro, wind and solar power. Complete reliance on renewable power is becoming the ultimate goal for most governments due to concerns regarding carbon emissions, pollution and global warming. Total renewable power would need to use recent emerging technology of large-scale BESS (Battery Energy Storage Systems) to store the energy when demand is low and release the energy when demand is high. There are already many of these innovative systems operating globally.

There are many variations of public charging stations that provide differing speeds of charging, with slow charging common for private households to more powerful charging stations sited on public roads and highways. As an example of how different chargers perform, the BMW i3 can charge 0–80% of the battery in around 30 minutes in DC rapid charging mode, whereas on a home charger, the charge time rises to about 8 hours. In contrast, the Tesla superchargers can supply up to 150 kW of charging, allowing a 300-mile charge in just under an hour. As battery technology advances, new models such as the Porsche Taycan uses 1000V batteries and will soon be able to accept charge at twice the Tesla rate.

ighlights

g is a single-phase plug that allows charging power levels of up to 7.4 . 0V, 32 A). This protocol is mainly used in EV's and is rare in Europe, being mainly confined to the Asian region. See figure 15.

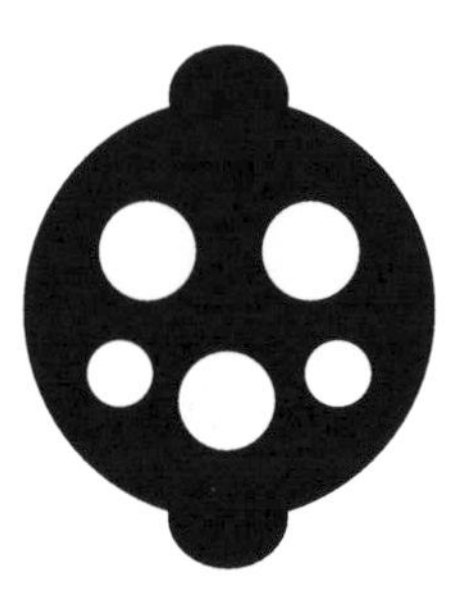

Figure 15 **Type 1 Connector**

Type 2

Although infrastructure for EV public charging points is expanding, they are still not common outside urban conurbations. Type two, affectionately known as slow to fast chargers, far outweigh any other type of charger in most countries The IEC 62196 Type 2 connector is used for charging electric cars predominantly in Europe. The connector is circular in shape, with a flattened top edge and originally specified for charging battery electric vehicles at between 3–50kW AC. See figure 16

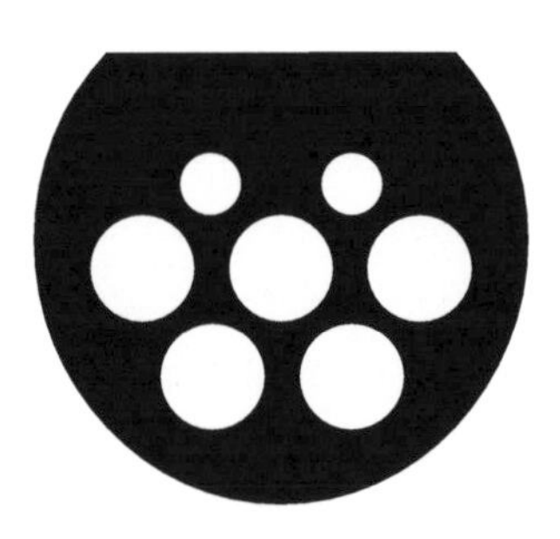

Figure 16 **Type 2 Connector**

Type 3

Finally, there is the growing expansion of Type 3 rapid or ultra-chargers that use DC power. Type 3 chargers can provide a full charge in 30 to 50 minutes and are often found on the main arterial routes and highways across countries, to facilitate long distance EV motoring. The very latest Type 3 chargers can charge in less than 30 minutes, with a planned 10-minute recharge in the not too distant future. They typically charge at rates from between 50kW to 150kW DC. See figure 14 for variants.

FACT

CCS is the European charging connector standard that combines Type 2 fast charging with DC rapid charging in one connector.

CONDUCTIVE COUPLING

Conductive coupling is where a plug and a socket physically lock together for transfer of power, is used to supply electricity for recharging most electric vehicles. Although there is now a battle between manufacturers on what connector is standard, much the same as VHS versus Betamax during the 70's. In Europe and the USA, the CCS (Combined Charging System) is the preferred plug amongst manufacturers, with the exception of Tesla. Although European Tesla Model 3's come with CCS connectors now. Japanese manufacturers opted to use the CHAdeMO system, a connector completely different in form and function to the CCS European standard. The main differences are that CCS is a dual mode plug that incorporates the Type 2 plug and DC rapid charger plug combined, whereas the Japanese must use adaptors or two sockets to perform the same functionality. Figure 14. highlights the difference in charging protocols. Nevertheless, CCS seems to be gaining momentum as the connector of choice due to its dual mode flexibility.

INDUCTIVE CHARGING

Another approach is inductive charging, using an inductive plate inserted into a slot in the car or one that comes in close contact with the inductive receptor. This technology is not new for low current applications. You probably use this type of charging regularly at home, either with your toothbrush or mobile phone.

Figure 17 **Mobile Inductive Charging**

There are two main types of conductive charging options: Static and mobile. Static conductive charging occurs when you are parked up, perhaps in a parking bay. Once activated in park mode, the car draws charge much the same as a toothbrush or mobile phone, from a transmitter plate in the road and receptor plate on your car. This method of charging is not as efficient as conductive charging, but the positive feature of this method is that there are no cables that can be damaged or vandalised, so it is ideal to install on a public road. Figure 14 highlights the concept of mobile inductive charging – on the go. An EV would effectively collect power on the move, from inductive rails or coils, embedded just below the roads surface.

VEHICLE TO GRID: UPLOADING AND GRID BUFFERING

There are a number of countries where EV owners are now contributing to support the electricity grid during peak load periods when the generation cost can be very high. In this situation, EV's can contribute energy to the grid and in most cases get paid for their contribution. The owner's vehicles can then be recharged in off-peak hours, typically overnight, or by a renewable energy source such as solar panels at lower rates, whilst absorbing excess night time electricity generation. In this situation, the cars batteries act as a distributed energy storage system to cushion power demand. Portugal's island of Porto Santo in the Atlantic Ocean now has a growing scheme in place to assist in effectively minimising pollution generated by its diesel generating plant, by buffering and feeding back energy required during peak times of the day from their EV.

CHARGING INFRASTRUCTURE

EV's have long faced two significant issues, lack of places to charge and limited range. With each new model launch and battery technology breakthrough, electric car ranges are increasing too. Although is the charging infrastructure there to support the increase in new EV's to support the forecast upsurge in electric car use and will the current charging infrastructure support your lifestyle with the charging network in place now? This section will hopefully address these important questions and more.

The majority of EV users plug in at work and home. Most EV users park their cars for hours on end either outside offices, factories or houses. This situation is the perfect opportunity to charge your EV battery, so the car is fully charged whenever and wherever you need it. A modern Type 2 7kW unit can charge a BMW i3 from 10% to full in less than four hours, so overnight or working day charging is perfect. If you're charging from a normal 13A domestic socket though, the charge cycle can take up to 12 hours.

This method of charging makes a 180 mile (290km) trip realistic in the very latest EV's and the best part is that it will only cost a few pounds/dollars in power versus £20/$16 or more in a petrol/gas or diesel vehicle. The main concern for most potential EV owners is when there is a need to travel further than the EV's range allows in one trip. What options are there to top up your battery when away from home? The good news is that in the UK for instance, there are now more than 14,000 public charging points with more than 25,000 connectors and better still, more than 2,500 of these are rapid chargers, where you can recharge your EV in roughly 30 minutes to 80% [23]. This scenario is similar in the USA, with more than 50,000 charging stations and 5,000 being rapid chargers. The downside is that many of these chargers are now pay as you use, although this still equates to roughly a third the cost of using an ICE v EV.

We have witnessed a significant change in public EV charger network infrastructure globally. Originally, there were disparate charge point operators and some car manufacturers, rolling out patchy charge point coverage, many in places that were hard to get to, or not on main highway routes. Fast forward a decade and the intervention of government and some manufacturers and the picture is completely different, helping establish contiguous coverage across European mainland, the UK and the USA.

With the advent of EV's now starting to become mainstream rather than a niche purchase, the charge-point networks need to start earning revenue to help pay for the huge investment of infrastructure roll-out and to pay for the electricity consumed. While not ideal, it does focus the EV users mind, with more drivers choosing to charge at home, due to the lower cost per electricity unit, especially if you charge on an off-peak tariff.

Figure 18 **Car Charging and Mobile Control Application**

Without exception, all new EV's now come with a mobile application that allows the driver to remotely manage the vehicle while it is parked on or off charge via Wi-Fi and 4G or 5G.

Some of the key features that are available for mobile remote monitoring and control are:

» Charging status, energy consumption and emissions whilst remote from the car
» Ability to remotely pre-heat or cool the vehicle prior to driving
» Ability to delay charging until cheaper off-peak tariff energy starts
» Check on charging and available range remotely
» Check current location of car, track journeys and previous trips too
» Monitors driving style with tips and shows how economically you have driven
» Will alert you when charged, if you stop for a coffee to top up for instance

6. CAN THE GRID SUPPORT FUTURE EV GROWTH?

It is not likely that EV's will create a power-demand disaster but could reshape the demand curve. EV's could soon face a different kind of gridlock, with the transport electrification accelerating, electricity suppliers and distributors must comprehend the possible impact of EVs on electricity demand.

But according to McKinsey (2018), their analysis suggests that projected growth in electric transport will not drive major growth in total electrical-grid power demand in the near to mid-term, therefore limiting the need for new electricity-generation capacity during that period [24]. Furthermore, most private EV users charge at night when electricity demand is at its lowest.

7. ICE VERSUS EV'S: THE PROS AND CONS

PRO: OPERATING COSTS ARE CHEAPER

Although petrol/gas prices remain affordable in the US, that is not the case generally in the rest of the world. An EV costs less money to run per mile/km on electricity than petrol/gas. Furthermore, an electric car generally costs less to maintain due to a fraction of the moving parts found in an ICE. Servicing an EV is a much simpler task, with few consumables such as oil changes to worry about, no spark plug changes and no transmission system worries.

CON: NEW EVS ARE EXPENSIVE

Although prices are in a downwards trend and will drop significantly over time, you will still be paying an up-front premium to buy, but will start recouping that on electricity costs and lower maintenance charges.

PRO: INCENTIVES CAN MAKE EVS MORE AFFORDABLE

In almost all countries where EV's are on sale, they are eligible for tax credits or government grants, either local or national. These cash incentives effectively can cut up to 25 percent off the total price of an EV, depending on which country and state that you reside.

CON: DRIVING AN EV MAY NOT BE AS GREEN AS YOU THINK – OR IS IT?

The local source of electricity governs an EVs overall effect on the environment. If you have your own stored solar panel source, then emissions are zero from source to road. At worst, if your source is coal powered electricity, because an EVs motor is on average 75% efficient, and a conventional petrol/gas powered car engine is at best 20% efficient, then an EV is still far greener and of course, there are no tailpipe emissions polluting our towns either. Consequently, EVs are generally responsible for far less pollution than conventional vehicles in everyday use.

PRO: PRICES OF USED EVS ARE CHEAP

At the moment, prices of used examples of EV's are looking great value for money. Although, as the myths about endurance of battery packs and the vehicles in general are busted, residuals on EV's are starting to rise, which is great news for owners of new EVs. Furthermore, early examples of EVs had less range than their new counterparts and tended to be driven fewer miles than the average ICE. This equates to less wear and tear, again, making used EV's great value for money.

CON: THERE IS A LACK OF EV CHOICE

The choice to consumers is certainly not as expansive as the range of petrol or diesel vehicles across the main EV markets. Although, in EV's main markets, there is a greater choice available to buyers than ever before. In China, there are more than 20 manufacturers and 35 models to choose from. While in Europe and the US, there is more than 30 new models available now, with more than 25 new models coming out year on year in the near future.

PRO: SOME EV'S CAN TRAVEL MORE THAN 200 MILES (320KM) ON A CHARGE

Battery technology is being driven by EV manufacturers to ensure practicality for both daily commuters and long-haul drivers. The Tesla Model S has a range up to 335 miles (539km) per charge in its long-range battery variant and its smaller Model 3 can attain up to 300 miles (482km). The Jaguar iPace has a top range of 240 miles (386km), while the new KIA e-Niro boasts an operating range of 300 miles (482km).

CON: ALL EV'S ARE SUBJECT TO RANGE LIMITATIONS

You will need to keep a check on the state-of-charge gauge no matter which EV you drive. Although, that's true for a conventionally powered car too. Older EV's generally travel 60-90 miles (96-145km) before needing a charge, although to cover the average commute, that range is fine. All EV's rated range can be negatively impacted by other factors. These include driving in extreme hot or cold weather that affects a battery's charge, and discharge and heater and air conditioning drain on power. Additionally, hard acceleration, poor maintenance and driving at higher speeds also impacts an EV's driving range.

PRO: EVS ARE FAST AND QUIET

An electric motor produces 100% of its available torque instantly, unlike a conventional combustion engine. This means that power is fed to the wheels immediately, allowing formula one style launches and great high-torque overtaking capabilities, with little noise. In fact, all that can be heard is normally the feint hum of the motor and tyre noise. Some new models have added sound tracks to the car that add to the drama and can only be heard inside, such as the Jaguar i-Pace. This is meant to add aural exhilaration to the silent driving experience.

CON: THERE ARE INADEQUATE PUBLIC CHARGING POINTS

Although infrastructure for EV public charging points is expanding, they are still not common outside urban conurbations. There are two distinct types of public charging points. Type 2 slow or semi fast chargers, far outweigh any other type of charger in most countries They are great for minor replenishment, taking between 3 to 10 hours to charge modern EV's to 80 percent charge. Finally, there is the growing breed of Type 3 rapid or ultra chargers. Type 3 chargers can provide a full charge in 30 to 50 minutes and are often found on the main arterial routes and highways across countries, to facilitate long distance EV motoring.

PRO: THE PLEASURE OF DRIVING A ZERO-EMISSION VEHICLE

EV's produce zero tailpipe emissions, unlike combustion engine powered vehicles. They don't emit greenhouse gases and air-bound pollutants, including oxides of nitrogen, carbon monoxide, formaldehyde, carbon dioxide, particulate matter and non-methane hydrocarbons into the atmosphere.

PRO: NO MORE WEEKLY PETROL OR GAS STATION VISITS

You will avoid the weekly visit to the petrol/gas station because most EVs are charged at home, mainly overnight. You may be able to secure a low night rate dependant on your electricity supplier. To fully charge an EV using standard Type 1 domestic current can take more than 24 hours, dependant on the battery pack size fitted to the model of EV. Most countries will subsidise a 230 or 110V Volt (country specific) Type 2 type charger box and this can dramatically reduce charge time to as little as four to five hours.

To recap: EV's have their pros and cons like most consumer goods, though, the Pros do seem to counter the Cons for most users:

PROS

- Quiet and fast
- No tailpipe emissions
- Low maintenance
- Cheap to run
- Growing reputation
- Safe and fun to drive
- Cheaper as company cars
- Hold their value
- Local and national government incentives and grants

CONS

- High purchase price… But this is offset by government incentives, grants and low running costs.
- Long charging times… But growing rapid charge network means ultra-fast recharging times available now.
- Lack of consumer choice… But this has already improved dramatically and will continue to expand.
- Limited range per charge… but again, rapid charge network means very fast recharging times now.

FACT

EV's generally have a lower centre of gravity than ice vehicles (due to floor mounted batteries), making them less susceptible to roll-over.

8. THE BUYERS DILEMMA

According to a study by McKinsey (2017), in the US, 30 percent of the vehicle buyers that McKinsey surveyed reported that they contemplated buying an EV, but only 3 percent actually did so [25]. The figures were comparable in Germany where 45 percent considered buying an EV, but only 4 percent did so. Though, in Norway roughly 22 percent of potential buyers eventually bought an EV, but this appeared to correlate directly to the generous government subsidies made available, the report stated. The survey was conducted online and involved 3,500 respondents from the US, Germany, and Norway. This data compares positively with widely held research that more than two thirds of people prefer the status quo in their life, rather than disrupt things with change.

By simply reading this guide and understanding its potential life changing outcome in both quality of life and helping the environment now and for future generations, then by default, you are probably one of the third of positive change agents in society.

McKinsey also argues that roughly 50 percent of potential car buyers are unaware of how EV's and associated infrastructure works. Although, by the end of this book, one would hope that you will be armed with a new found comprehensive knowledge of how EV's and its supporting infrastructure works both for you and the environment.

We know that over the next decade, battery and component parts will continue to fall in price as we head towards price parity, according to Deloitte research [26], by 2024 to 2025 (see figure 19). It is also a fact that the charging infrastructure is expanding faster than the number of vehicles being registered on the roads, particularly in the US and UK. Range anxiety, the phenomenon of not knowing if your EV will make it to the next charging point, is clearly becoming less of an issue as most countries are now at, or close to contiguous charge point coverage on all main routes.

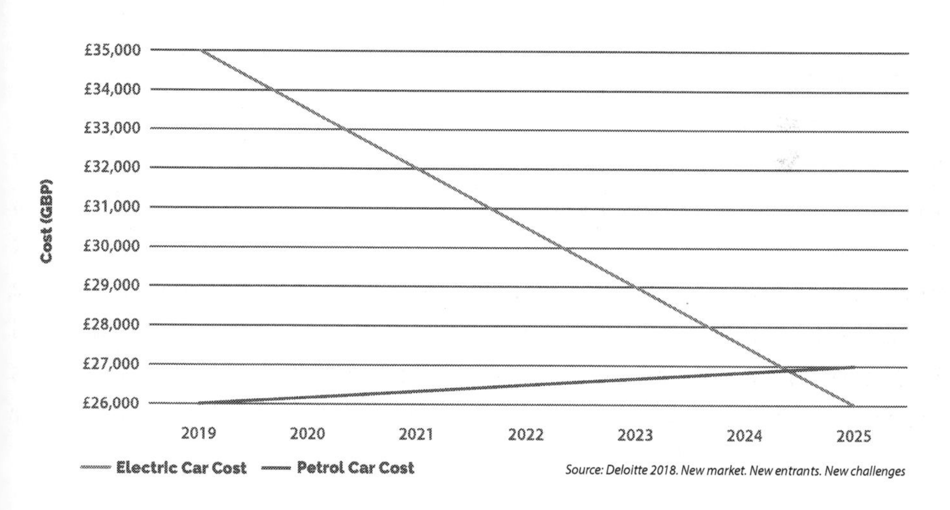

Figure 19 **Electric Car Versus Petrol Car - Price Parity**

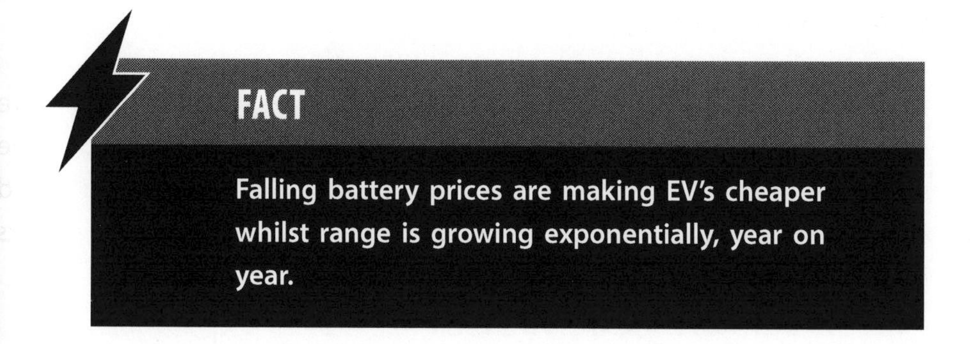

9. BUYING GUIDE

Is it worth getting an electric car right now? It is clear that they are more expensive than their combustion engined distant relatives at the moment, but aside from price, an EV will fulfil your same requirements as petrol/gas or diesel vehicle, but with much lower fuel, maintenance and lower running costs. You will of course need to evaluate your average journey lengths. Are most of your trips lower than the range of your EV shortlist? Will you have opportunities on the way to recharge if your journey exceeds the range of your EV of choice? If the answer is yes, then the next chapter provides an up to date list of new and near future BEV's.

Like all manner of new technology, BEV's are improving on an almost daily basis, but that doesn't mean you should put off the inevitable right now. If it truly works for you, there has never been a better time to buy an EV. Development in technology now offers much faster recharge times, long haul service stops with ultra and rapid DC charging. So, you can charge the car, have a welcome rest break and coffee, then both you and the car will be refreshed in about 30-40 minutes. Range is much less of an issue now, as the rapidly expanding charging infrastructure is growing by the week, this means low to no range anxiety Alternatively, if you opt to delay, you could be driving ever more innovative EV designs, from most of the large car manufacturers, as they strive to either catch up or even beat the competition.

IS NOW THE THE RIGHT TIME TO OWN AN EV?

It is surprisingly familiar buying a used EV. Much the same as purchasing, hiring or renting a conventional combustion engined car, and some aspects are even better. Less pollution and no tailpipe emissions, easy automatic (single speed) transmissions systems, better performance and less brake wear, due to the use of regenerative braking most of the time. What's not to like? Whether you buy new or used.

Most new EVs come with at least 3 years warranty on the car and 8-10 years on the battery. Even if the car is used and out of warranty, if things go wrong, repairs should cost no more than a conventional car. Although, if buying a used EV, remember if possible, try to buy through a reputable dealer and get a good warranty included.

FACT

EV tax breaks and government grants benefit drivers in all income levels.

ASTON MARTIN Rapide E

Australia, Canada, Europe, New Zealand and USA

MANUFACTURERS RANGE

200 miles – 320km (est.)

HOME CHARGE TIME

8 hours (est.)

RAPID CHARGE TIME

50 minutes to 80% (est.)

For Aston Martin, this new version of the Rapide is a giant leap into the future and the basis for many future models. It comes with a range of 310 miles and a restricted top speed of 155mph. The only downside being its £250,000 price tag.

The 625bhp Rapide E is a true 4 door, 4 seat super car. Its 0-62mph (100kph) time of less than 3 seconds proves it's no slouch too. The huge power train has evolved through a partnership with F1 specialist, Williams Advanced Engineering and its partner - Unipart.

There is no doubt that Aston Martin are registering their intent that they are here for the long-run, by preparing and positioning themselves for a total electrification strategy into the next decade.

PROS

» Spacious 4 seat, 4 door sports saloon
» Looks sublime and interior superb
» Prestigious badge
» Ultra-luxury EV saloon

CONS

» Hugely expensive
» Pricier than petrol version
» Less range than much cheaper rivals
» Expect high depreciation

PRICE	£250,000 (Est) - $300,000 USD
ON SALE	Now

AUDI E-Tron

Australia, Canada, Europe, New Zealand and USA

MANUFACTURERS RANGE

240 miles – 385km

HOME CHARGE TIME

7 hours (est.) to 80%

RAPID CHARGE TIME

50 minutes to 80% (est.)

Following years of trade show prototypes and concepts, the first all-electric Audi is finally here.

The Audi E-Tron is the company's first road going production BEV. Its rival's number the Jaguar iPace, Mercedes EQC and Tesla Model X.

It's charging capabilities are groundbreaking and it's confirmed that when available, its batteries will be able to accept a 300kW charge, but for now most countries will have to accept rapid charging between 50kW and 150kW maximum.

Performance is very good. Its maximum usable range is 240miles (385km) and can be rapid charged on a 50kW station from 10% to 80% in just 50 minutes. The E-Tron's motor pumps out a whopping 402 bhp to all four wheels and it can accelerate to 62mph (100kph) in just 5.7 seconds. A great all-round BEV SUV.

PROS

- Spacious 4 seat, 4 door SUV
- Looks substantial
- Prestigious build and badge

CONS

- Expensive
- Higher price than petrol version
- Less range than cheaper rivals

PRICE £72,000 - $76,000 USD

ON SALE Now

BMW i3

Australia, Canada, Europe, New Zealand and USA

MANUFACTURERS RANGE
150 miles – 240 km

HOME CHARGE TIME
3.5 hours (est.)

RAPID CHARGE TIME
40 minutes to 80% (est.)

Now in its 3rd generation guise and now only available as a BEV, the BMW i3 is still turning heads and admiring glances through the window to witness a miracle in modern design and packaging. Its Tardis like interior belies its short stumpy exterior.

The i3 is classed as a luxury compact 5 door hatchback and is now equipped with larger batteries, it has the range to match its price (well almost). It's wholly justified by its bespoke carbon fibre all in one body and spaceframe, making it one of the lightest, strongest 4 door saloon bodies in the world.

Coming with a 168bhp motor driving the rear wheels, 80% charge in just 40 minutes and a 0-62m (100km) acceleration time of 6.7s, it doesn't disappoint.

PROS

» Roomy 4 seat, 4 door small hatchback
» Space age looks
» Great build quality
» Prestigious badge

CONS

» Expensive for small hatchback
» Quirky looks for some
» Pricier than petrol version
» Less range than cheaper rivals

PRICE £35,000 UK - $45,000 USD
ON SALE Now

CHEVROLET BOLT

USA and Canada.

MANUFACTURERS RANGE

240 miles – 340km

HOME CHARGE TIME

9 hours (est.)

RAPID CHARGE TIME

60 minutes to 80% (est.)

The Chevy Bolt is Americas answer to the BMW i3. Although it differs in drive train. The Bolt is front wheel drive, while the i3 is classic BMW rear wheel drive.

The bolt has a very reasonable range of 240 miles (340km) and can accelerate from a standstill from 0-62mph (100kph) in just under 7 seconds using its powerful 200bhp motor. Many commentators still think that from a distance, side view the Chevy looks like a 15-year old Mercedes A class, but overall the car represents good value for money considering its great range.

From 10% to 80%, using a 50kW fast charger, the Bolt charges in 60 minutes. Overall, the lack of basic tech, such as GPS Satellite Navigation, lets down an otherwise excellent compact 5 seat, 4 door hatchback.

PROS

» Spacious 5 seat, 4 door hatchback
» Looks great
» Good build quality
» Good range

CONS

» Expensive
» Pricier than petrol equivalent
» Less range than cheaper rivals

PRICE $36,500 USD
ON SALE Now

CITROEN C Zero

Australia, Europe, and New Zealand

MANUFACTURERS RANGE
90 miles – 144km (est.)

HOME CHARGE TIME
5 hours (est.)

RAPID CHARGE TIME
N/A

The Citroen C-Zero is on the surface, a brilliantly packaged BEV wrapped in a traditional ICE body, shared with Peugeot and Mitsubishi.

Although when you inspect the performance and specification in detail, it's priced competitively for good reason. Its maximum range (in summer) is just 90 miles, but in reality, will be closer to 65miles. Its 0-62mph (100kph) acceleration is just 16 seconds and a comes with a top speed of 83mph (133kmh).

An absence of DC rapid charging means that it charges only on a domestic mains socket or a 32A fast charger, the latter taking about 3 hours (10% to 80%).

PRICE £ 20,500 (not US)
ON SALE Now

Despite the relatively low cost, the Citroen C zero is not without its flaws and will eventually be replaced by Peugeots upcoming models, judging by recent and near future arrivals, nipping at its heels.

PROS

» Compact 4 seat, 4 door mini hatch
» Looks good
» Great for around town use
» Easy to park

CONS

» Limited interior space
» Pricier than petrol version
» Poor range

HYUNDAI KONA Electric

Australia, Canada, Europe, New Zealand and USA

MANUFACTURERS RANGE
190 miles – 305 km

HOME CHARGE TIME
5 hours (est.)

RAPID CHARGE TIME
40 minutes to 80% (est.)

The Kona Electric is Hyundai's version of its sister company, KIA's E-Niro. The Kona delivers pretty good power, impressive fit and finish and provides a very comfortable ride.

The Kona Electric is based on the same platform and body shell as the rest of the Kona range and therefore is compromised on space and overall performance due to its extra weight due to its battery packs.

The Kona's standard 134bhp electric moto delivers a range of 190 miles (305 km) and 0-62mph (100kph) performance of 8.6 seconds.

Surprisingly, this Korean built car comes equipped with the European standard CCS charging socket, capable of being used with a Type 2 charging plug and also combined CCS plugs for rapid DC charging.

PROS

» Spacious 4 seat, 4 door hatchback
» Practical exterior design
» Well-equipped and good build quality
» Good to drive

CONS

» Moderately expensive
» Unexciting interior
» More expensive than standard Kona variants
» Range not as good as KIA e-Niro

PRICE £30,500 - $37,000 USD
ON SALE Now

JAGUAR iPace

Australia, Canada, Europe, New Zealand and USA

MANUFACTURERS RANGE
255 miles – 410km

HOME CHARGE TIME
12 hours (est.)

RAPID CHARGE TIME
70 minutes to 80% (est.)

Jaguar's first fully electric car has taken the motoring world by storm and come as a shock to many established car companies, such as Tesla. Deservedly winning many global and national car of the year awards in 2018 and 2019.

But let's see exactly what makes the first all-electric Jaguar so good.

The iPace has got reasonable range at 255miles (410km) maximum compared with its rivals and the fit and finish is up there with the very best current crop of BEV's.

The iPace continues the brands focus on performance and value. It can rapid charge from 10% to 80% in 70 minutes and its 396bhp engine powers the iPace from 0-62 mph (100kph) in just 4.5s. What's not to like at this price?

PROS

» Spacious 5 seat, 5 door mid-SUV
» Excellent design and distinct looks
» Great real-world range
» Prestigious badge

CONS

» Very expensive
» Firm suspension
» Long home charge time

PRICE £61,000 - $70,000 USD
ON SALE Now

KIA E-Niro

Australia, Canada, Europe, New Zealand and USA

MANUFACTURERS RANGE
260 miles – 420km

HOME CHARGE TIME
12 hours (est.)

RAPID CHARGE TIME
80 minutes to 80% (est.)

If the KIA E-Niro looks familiar, it's because the Hyundai Kona electric is essentially the same car and set-up, with different clothing. Of course, there are many visual differences, but both cars are pioneers in virtually achieving the holy grail of EV's, by reducing the price of a family electric SUV, whilst at the same time increasing the range to almost ICE like ranges.

Using the same powertrain as the Kona, the E-Niro is powered by a 201bhp motor fed by a 64kW battery pack, powering the E-Niro from 0-62mph (100kph) in a brisk 7.8s.

Charging the E-Niro via a 50kW rapid charger from 10% to 80% is achieved in 80 minutes due to the large capacity power pack, providing a range of approximately 260 miles (420km). All in all, this is one of the best, accessible long-range electric family SUVs. Fit and finish is very good, but interior is still a little austere.

PROS

» Spacious 5 seat, 5 door family SUV
» Great range
» Practical and comfortable
» Standard high level of equipment

CONS

» Moderately expensive
» OK styling
» Servicing intervals quite short
» Moderate performance

PRICE £36,500 - $40,000 USD
ON SALE Now

KIA SOUL EV

Europe only

MANUFACTURERS RANGE
280 miles – 450 km

HOME CHARGE TIME
5 hours (est.)

RAPID CHARGE TIME
50 minutes to 80% (est.)

Unlike most mainstream manufacturers, this is already KIA's second fully electric car.

The quirky but practical KIA Soul EV is a cross over model and because of its boxy shape, is immensely practical and spacious, especially load space via the rear hatch.

Unlike their other BEV – the KIA E-Niro, the Soul EV will only be on sale in Europe. There will be a choice of two electric drive trains:

- 64kW battery with a 201bhp motor. 280mile (450km) range – 0-62mph 7.6s
- 39kW battery with a 134bhp motor.172mile (276km) range – 0-62mph 8.9s

PROS

- Small but fairly spacious family car
- Unique styling
- Comes with high level of equipment
- Smooth drive

CONS

- Priced high compared to standard car
- Plain interior
- Good range for class

PRICE £26,500 - £32,000 USD
ON SALE Now

MERCEDES EQC

Australia, Canada, Europe, New Zealand and USA

MANUFACTURERS RANGE
250 miles – 402km

HOME CHARGE TIME
8 hours (est.)

RAPID CHARGE TIME
40 minutes to 80% (est.)

The EQC is Mercedes first entry into their new range of electric cars and purpose-built platforms. The EQC is a direct rival to the Jaguar iPace and the Audi E-Tron, all in the luxury SUV class, but all much smaller in size to the Tesla Model X.

The EQC comes with an impressive 402bhp motor and is powered by a large 80kw battery pack. Its range is a very impressive real world 250 miles (402km) and charging using a 100kw ultra charger from 10% charge, takes a mere 40 minutes.

The EQC is 4-wheel drive and produces a rapid 0-62mph (100kph) acceleration time of just 5 seconds, which considering its 2.5 tonne weight is quite impressive.

PROS

» Spacious 4 seat, 5 door exec SUV
» All wheel-drive
» Looks great
» Prestigious badge
» Comes with good standard of kit

CONS

» Expensive
» Much pricier than petrol equivalents
» Less range than cheaper rivals or direct competitors

PRICE £70,000 – $79,000 USD
ON SALE Now

NISSAN LEAF

Australia, Canada, Europe, New Zealand and USA

MANUFACTURERS RANGE
160 mile – 260 km

HOME CHARGE TIME
8 hours (est.)

RAPID CHARGE TIME
45 minutes to 80% (est.)

Now in its 2nd generation guise, the Leaf has been around since 2010, yet it seems that it's been on our roads forever.

A true EV pioneer that has set the bar for every EV launch since BEV's became a common sight on our roads. In its new modern format, the Leaf comfortably travels 160 miles (257km) or more on one full charge and can be charged from 10% to 80% using a rapid charger in 40 minutes.

Like most new EV's, it is equipped with regenerative braking that provides remarkable one pedal driving. With plenty of room for 4 adults and a decent boot or trunk that is larger than a VW Golf, this car is now one of the best BEV's for range and value bar none. If you like the undistinctive styling but appreciate its great driving abilities and roomy family friendly interior, then this could be for you.

PROS

» Roomy small family saloon
» Competitively priced
» Self-driving add-on soon
» Proven reliability

CONS

» Hard suspension
» Staid and dated interior
» Rivals starting to overtake on price, performance and quality

PRICE £26,700 - $30,000 USD
ON SALE Now

PEUGEOT e208

Australia, Canada, Europe, New Zealand

MANUFACTURERS RANGE

200 miles – 322km (est.)

HOME CHARGE TIME

5 hours (est.)

RAPID CHARGE TIME

30 minutes to 80%

The new Peugeot e208 made its debut at the 2019 Geneva motor show.

The brands latest BEV is based on the same platform as the entire range of new 208's and is being treated as simply an additional engine option.

The e208 is fitted with a new 134bhp motor and 50kw battery, providing a range of just over 200 miles (321km) on a full charge and via a 100kW ultra charger, can be topped up in 30 minutes to 80%.

The battery pack has been positioned in such a way that it does not impact on the new e208 range interior space. Additionally, the latest tech is available, including lane keep assist and traffic sign recognition. All in all, this is a great package, with latest tech and decent range.

PROS

- Compact 4 seat, 5 door-hatch
- New design looks modern yet familiar
- Great for medium range commutes
- Easy to park

CONS

- Pricier than petrol version
- Range less than its competitors
- Revamped 208 ICE interior

PRICE £ 27,000 – ($32,000) USD

ON SALE Now

RENAULT ZOE

Australia, Europe, and New Zealand

MANUFACTURERS RANGE
170 miles – 275km

HOME CHARGE TIME
6 hours (est.)

RAPID CHARGE TIME
95 minutes to 80%

When Renault launched the first Zoe, they attempted to address two common concerns that potential buyers had regarding electric cars: how far the cars can travel on one charge and the high price perception of EV's versus conventional ICE powered cars. Renault listened to feedback from customers and set about introducing the ZOE that addressed these two main issues.

The latest ZOE incarnation provides a lower purchase price for a compact family EV hatch than its competition, by leasing the battery, Renault eliminated user fears of early battery degradation. Secondly, the battery pack was redesigned to squeeze in the maximum size that current technology could provide.

The net result is a great looking, practical family hatch, capable of travelling 170 miles (275km) on a full charge and by using a rapid charger, it can be topped up to 80% in 95 minutes. Its 107bhp motor powers the Zoe from 0-62 mph (100kph) in 8.7s to 13.1s depending on model.

PROS

» Modern practical design
» Reasonable range
» Battery lease reduces purchase price

CONS

» Mediocre handling
» Battery lease adds to monthly cost

PRICE £21,500 - $26,000 USD
ON SALE Now

SMART EQ ForFour

Australia, Canada, Europe, New Zealand and USA

MANUFACTURERS RANGE
65 miles – 104 km

HOME CHARGE TIME
6 hours (est.)

RAPID CHARGE TIME
40 minutes to 80% (est.)

The smart range has always produced desirable mini 2 seat cars and more latterly ICE versions that evolved into 4 door, 4 seat cars, at a price that far exceeds their true value, preferring form over practicality. This EQ ForFour version is no exception.

Although the four seat EV version in Smarts range is slightly longer, it's still easy to park and like its two-seat brother, there is no question that the EQ ForFour is going to be extremely cheap to run. The Smart 4-seater EV is just large enough to be classed as a small family car, (with very young children).

Slightly longer than its smaller brother, the ForTwo, this car is a great entry into the EV world for young growing families. Performance is not bad, with a 0-62mph (100kph) time of 12.7s and maximum range from its 16.7kw battery pack on a full charge is up to 65miles (104km).

PROS

» Small 4 seat, 4 door mini hatch
» Looks good
» Good range of standard equipment
» Prestige - Part of the Mercedes group

CONS

» Pricier than petrol - gas version
» Tiny range versus cheaper rivals
» Firm ride
» Small boot (Trunk)

PRICE £21,600 - $26,000 USD
ON SALE Now

SMART EQ ForTwo

Australia, Europe, and New Zealand

MANUFACTURERS RANGE
70 miles – 112km

HOME CHARGE TIME
6 hours (est.)

RAPID CHARGE TIME
40 minutes to 80% (est.)

If you are looking for a two seat second car for town use or need a low range urban run-about for very short commutes, then the Smart EQ ForTwo could be the car for you.

The smart range has produced many desirable mini 2 seat cars and more latterly 4 door, 4 seat cars, at a price that far exceeds their true value, preferring form over practicality.

But there is no point ignoring its diminutive size, making it ideal for parking, and there is no question that the EQ ForTwo is going to be incredibly cheap to run.

It's priced similarly to the Renault Zoe, which is a much larger car, although the ForTwo is not aimed at families and it's an unfair comparison. Performance is not too bad, with a 0-62mph (100kph) time of 11.5s and maximum range from its 16.7kw battery pack on a full charge is up to 70miles (112km).

PROS

» Tiny 2 seat, 2 door micro hatch
» Very good turning circle
» Quick charge time
» Prestige - Part of the Mercedes group

CONS

» Pricier than petrol - gas version
» Short range compared to rivals
» Firm ride
» Very small boot (Trunk)

PRICE £21,195 - $27,000 USD
ON SALE Now

TESLA model 3

Australia, Canada, Europe, New Zealand and USA

MANUFACTURERS RANGE
From 220 miles – 354km

HOME CHARGE TIME
14 hours (est.)

RAPID CHARGE TIME
50 minutes to 80% (est.)

The Tesla model 3 has had more pre-launch hype than any other BEV in the company's history. Touted as the peoples EV due to its entry price and range, the price has gradually crept up prior to launch and the range hasn't materialised to match Tesla's original fanfare.

Nevertheless, following production delays, the model 3 has evolved into a superb, mid-size family car, with a very usable 220 miles (354km) range at entry level. It is a car that can be used for all trips and given the top level 3's generous range, it is a true long-distance BEV for the well-heeled masses. Acceleration of 3.7s – 0-62mph (100kph) and a top speed of 139mph (224kph), makes the model 3 a compelling proposition...until the model Y arrives.

The model 3 will undoubtedly be Tesla's best-selling car, that is until the company launches the soon to be released compact SUV – Model Y. The model 3 is the best quality car that Tesla has produced yet.

PROS
- » 5 seat, 4 door Saloon
- » Prestigious EV badge
- » Awesome performance

CONS
- » Expensive
- » Not much rear room space
- » Interior not up to European standards

PRICE £40,000+ $38,000 USD
ON SALE Now

TESLA Model S

Australia, Canada, Europe, New Zealand and USA

MANUFACTURERS RANGE
240+ miles – 386km

HOME CHARGE TIME
13 hours (est.)

RAPID CHARGE TIME
40 minutes to 80% (est.)

The Tesla model S was a landmark in both the company's history and to the world of electric vehicles as a whole. It was the first BEV in the world that you could genuinely use for all trips, given its generous range, passenger space and astounding performance.

Launched in its original form in 2014, the model S is now in its second-generation format and refresh. The only downside of this impressive BEV is its price.

The Tesla Model S in its entry guise, has a 75kw battery pack feeding 310bhp of motor power driving all 4 wheels. Its top speed is 155mph (250kph), sprinting 0-62mph (100kph) in 5.9s.

Quality was initially very poor for a car priced in the top tier of luxury cars, but this has improved enormously, although fit and finish is still not up to European standards.

PROS

» Great range – similar to petrol/gas
» Spacious 4 seat, 4 door saloon/sedan
» Lightning fast
» Futuristic tech and standard kit good

CONS

» Very expensive
» Cost, much more than petrol luxury cars
» Huge weight
» Quality not up to European standards

PRICE £88,000+ - $75,000+ USD
ON SALE Now

TESLA Model X

Australia, Canada, Europe, New Zealand and USA

MANUFACTURERS RANGE
280 miles – 450km

HOME CHARGE TIME
13 hours (est.)

RAPID CHARGE TIME
40 minutes to 80% (est.)

The Tesla Model X was a brave move by this pioneering company and is a world apart from their very first BEV, the Tesla Roadster, in the form of the British built Lotus Elise.

Fast forward to the four-wheel drive Tesla X. It's one of the largest family SUV's in any form and this beast outperforms most sports cars in a straight line, accelerating 0-62mph (0-100kph) in just 3.0s.

Its top of the range P100D model, covers up to 280 miles (450km) on a full charge. Using one of Tesla's superchargers, you can charge the model X from 10% to 80% in just 30 minutes and using a 32A home charger, it will take about 11 hours to perform the same task. Top speed is a heady 153mph (246kph).

The X comes loaded with high tech, such as autonomous driving and it's this equipment list that sets it apart from most of its rivals.

PROS

- Great range – similar to petrol/gas
- Huge SUV – versatile interior layout
- Great real-world range
- Self-driving technology available

CONS

- Very expensive
- Firm suspension
- Huge weight
- Interior not to European standards

PRICE £91,000+ - $80,000+ USD
ON SALE Now

VOLKSWAGEN e-UP

Australia, Europe, and New Zealand

MANUFACTURERS RANGE
79 miles – 127km

HOME CHARGE TIME
6 hours (est.)

RAPID CHARGE TIME
40 minutes to 80% (est.)

Volkswagen have played safe in order to get a foot in the proverbial BEV door and unfortunately this has made their entry level small hatch twice as expensive as their petrol/gas driven counterpart and more expensive than the larger and longer-range Renault Zoe.

Although, for a small hatch that has been electrified, the e-UP is one of the best available in this class, if price is not a barrier, build quality is excellent. Through ingenious packaging, VW have created a BEV hatch that is fun to drive and practical for short commutes and shopping runs. Its range realistically prevents longer journeys.

Though, it can be rapid charged in 30 minutes and its 80bhp engine driving the front wheels can take you from 0-62mph (100kph) in a sedentary 12.4 seconds.

PROS

» 4 seat, 5 door mini hatch
» Identical to petrol/gas version outside
» Perky electric drive
» Perfect for use around town

CONS

» Much pricier than petrol/gas version
» Less range than cheaper rivals
» Newer rivals offer better tech now

PRICE £24,600 – (not US)
ON SALE Now

VOLKSWAGEN e-Golf

Australia, Canada, Europe, New Zealand and USA

MANUFACTURERS RANGE
140 miles – 225km

HOME CHARGE TIME
6 hours (est.)

RAPID CHARGE TIME
60 minutes to 80% (est.)

Volkswagen played safe with their first full production e-Golf by using the familiar body of the million selling Golf. The e-Golf concept was realised back in 2011 when VW started testing 500 prototypes in the field. But it wasn't until 2014 that the e-Golf in its current guise was launched in Germany.

Its 134bhp motor is capable of being rapid charged from 10% to 80% in 60 minutes and power the e-Golf from 0-62mph (0-100kph) in 9.6 seconds. Not as refined or revolutionary as VWs new range of electric cars due out soon, it has succeeded in providing a welcome substitute for the company to satisfy their growing army of EV customers. Its range is limited by today's standards and because the VW's battery pack is only cooled passively, it limits the continuous speed and journeys that you can partake as an owner.

Good for urban use and low mileage trips, so long as range is not an issue.

PROS

» Good 5 seat, 5 door-hatch
» Looks familiar – same as regular Golf
» Not bad range – great for small commute

CONS

» Quite expensive considering low range
» Firm ride
» Small boot/trunk compared to standard car

PRICE £30,000 - $35,000 USD
ON SALE Now

BMW i4

Australia, Canada, Europe, New Zealand and USA

MANUFACTURERS RANGE
340 miles – 547km (est.)

HOME CHARGE TIME
10 hours (est.)

RAPID CHARGE TIME
60 minutes to 80% (est.)

The all new BMW i4 sets the scene for the next generation of all EVs in BMW's future strategy. Although the new i4 adopts BMW's familiar family external deign traits, its enlarged outline kidney grill performs a purely cosmetic role with what looks like slim air intakes top and bottom of the internal grill itself, aiding passive electronic and battery cooling. The main body shell is based on the large gran coupe using their existing CLAR platform. This helps control costs on low volume models.

The i4 claims a top speed of 124mph (200kph) and 0-62mph (100kph) of just 4 seconds. Depending on model, the i4 can travel up to 340miles (547km) on a full charge and its charge rate from 10% to 80% using the new 100kW rapid chargers is an impressive 60 minutes. This new BEV is a gamechanger for BMW.

PROS

» Spacious 5 seat, 5 door compact executive
» Futuristic looks
» Very high tech – autonomous driving
» Good range

CONS

» Expensive
» Range similar to cheaper rivals

PRICE £55,000 - $65,000 US (Est)
ON SALE Late 2020

DS3 E-TENSE

Australia, Canada, Europe, New Zealand and USA

MANUFACTURERS RANGE
180 miles (est.) – 289km

HOME CHARGE TIME
6 hours (est.)

RAPID CHARGE TIME
40 minutes to 80% (est.)

DS is the French luxury brand and this new model is a perfect example of how their innovative design principles carry over to their first luxury BEV, the DS3 E-Tense.

Built on a modified platform of the DS3 range, they share the same underpinnings and infotainment systems as their ICE powered relatives.

The DS3 E-Tense is driven by a 136bhp electric motor and powered by a floor mounted 50kW battery pack.

This DS3 E-Tense powertrain will provide a 0-62mph (100kph) time of 8.7 seconds and an overall driving range estimated at 180miles (289km).

The DS3 E-Tense is capable of being charged with the new 100kW rapid chargers from 10% to 80% in just 40 minutes.

PROS

» Spacious 5 seat, 5 door family hatch
» Looks good
» Can be used on latest 100kW rapid chargers

CONS

» Fairly expensive for range
» Average performance
» Less range than cheaper rivals

PRICE £30,000 (Est) - $36,000 US (Est.)
ON SALE Early 2020

HONDA E

Australia, Canada, Europe, New Zealand and USA

MANUFACTURERS RANGE
120 miles – 194km (est.)

HOME CHARGE TIME
5 hours (est.)

RAPID CHARGE TIME
30 minutes to 80% (est.)

Honda has a peerless reputation of not only being an innovator of some of the worlds best everyday road cars, but in the top tier of F1 design and innovation too.

Enter the Honda E. This is a purpose designed 4 door, 4 seat 5 door mini hatch and marketed as an everyday urban runaround, with only 120mile (194km) maximum range and priced at an estimated £25,000 or $32,000.

The exterior is very cute, almost Golf Mk1esque. But the inside is a tour de force of minimalism and quirkiness all wrapped up in one superb futuristic design package. Despite the tiny dimensions, inside and out, this great city car is destined to be a hit and a cult car for years to come.

PROS

- Compact 2 door, 4 seat hatch
- Looks cute
- Built only as an EV with useful tech

CONS

- Average range
- Pricier than competitors
- Minimalist interior

PRICE £25,000 - $30,000 USD
ON SALE Late 2019

LAGONDA SUV

Australia, Canada, Europe, New Zealand and USA

MANUFACTURERS RANGE
350 miles – 547km (est.)

HOME CHARGE TIME
8 hours (est.)

RAPID CHARGE TIME
60 minutes to 80% (est.)

The all new Lagonda SUV will be the first EV from the luxury arm of Aston Martin and will be built in the company's new plant in St Athens, South Wales. Aston Martin has relaunched the Lagonda range as the first of many all-electric luxury sports cars and SUV's.

This powertrain and battery will feature in their first luxurious all-wheel drive, all terrain EV SUV, and this futuristic technology will feature in all upcoming models in the Lagonda portfolio. With a range of 350 miles (547km) on a full charge and capable of receiving a charge to 80% in just 60 minutes, this car will be a serious contender for ICE luxury saloons and SUV's.

The Lagonda SUV performance specification includes a 0-62mph (100kph) time of 3 seconds. The motoring world waits with bated breath to witness this pioneering EV SUV.

PROS

» Spacious 5 seat Luxury SUV
» Futuristic
» Prestigious badge
» All wheel drive

CONS

» Hugely expensive
» Less range than cheaper rivals
» May experience large depreciation

PRICE £300,000 (Est) - $380,000 USD
ON SALE Late 2020

MG ZS EV

Australia, Canada, Europe, New Zealand and USA

MANUFACTURERS RANGE
200 miles – 321km (est.)

HOME CHARGE TIME
6 hours (est.)

RAPID CHARGE TIME
40 minutes to 80% (est.)

The MG ZS EV is the first in a new line of BEV's from the Chinese owned MG brand. First revealed in Shanghai in 2018, this model is simply a battery powered version of its small family SUV hatchback.

The powertrain consists of a 148bhp motor driving the front wheels and its battery pack is estimated to provide an estimated 200mile (321km) range on a single charge. Acceleration is 0-60mph (1000kph) in 9.0s.

It is designed to be charged with the latest 100kW rapid chargers and has a capability of taking a full charge from 10% to 80% in a little over 40 minutes.

The launch price is estimated to be £25,000 ($30,000 USD) and will face rivals such as the new Renault Zoe and the Vauxhall Corsa electric.

PROS

» Compact 5 seat, 5 door SUV
» Same looks as ZS petrol/gas
» Good performance for price

CONS

» 2-wheel drive only
» Pricier than petrol version
» Low range compared with rivals

PRICE £25,000 (Est) - $30,000 USD
ON SALE 2020

MINI ELECTRIC

Australia, Canada, Europe, New Zealand and USA

MANUFACTURERS RANGE
190 miles – 305km (est.)

HOME CHARGE TIME
5 hours (est.)

RAPID CHARGE TIME
45 minutes to 80% (est.)

The legendary Mini reached its 60th birthday in 2019 and what better tribute than to release the much anticipated all electric Mini in its 61st year. The new Mini BEV will be the brands most revolutionary model, since the company tested early BEV versions of the mini way back in 2007. The main difference now is that the new Mini EV will be its first mass produced BEV and will be built in the UK at the brands Oxford HQ in Cowley. At launch, the company only has plans for a 3-door version, although it is expected that the Mini EV platform will carry through to all models eventually.

The new Mini EV is expected to use BMW's tried and tested 32kW battery pack powering the motor to drive through the front wheels. This provides the Mini EV with a 0-62mph (100kph) time of 6.8s. Rapid charge time will be just 45 minutes. Specifications will be high, and prices should be lower than the BMW i3.

PROS
- Compact Mini hatch
- Looks great
- Good range for a compact hatch
- Great performance

CONS
- Expensive
- Marginally less space than normal mini

PRICE £24,500 (Est) - $30,000 USD
ON SALE 2020

PEUGEOT ion

Australia, Canada, Europe, New Zealand and USA

MANUFACTURERS RANGE
90 miles (78km) est.

HOME CHARGE TIME
5 hours (est.)

RAPID CHARGE TIME
30 minutes to 80% (est.)

Making its debut at the Geneva Motor Show in 2019, Peugeot has shown its commitment to commencing complete electrification across its whole range by 2023.

This car is Peugeot's first EV to be built from the ground up, using a brand-new platform – not to be confused with the outgoing version ion.

This new model boasts a 30kW battery and is capable of producing real world range of just over 90 miles (144 km).

The battery is covered for 100,000 miles (169,935 km) and the car supports up to 100kw rapid charging, meaning that you can charge this cute new mini hatch on an ultra charger in under 30 minutes to 80%.

PROS

» Super mini 4 seat, 3 door-hatch
» Looks great
» High tech interior

CONS

» Expensive for a small hatch
» Less range than cheaper rivals
» Limited space (est.)

PRICE £26,000 (Est) - $29,000 USD
ON SALE 2020

PININFARINA Battista

Australia, Canada, Europe, New Zealand and USA

MANUFACTURERS RANGE
280 miles – 450km (est.)

HOME CHARGE TIME
12 hours (est.)

RAPID CHARGE TIME
90 minutes to 80% (est.)

The preserve of oligarchs and other billionaires, the new Pininfarina Battista is a tour de force of all that's best. If you have the money and crave complete individuality, then this is the car for you. Named after the founder and designer of the car, Battista Pininfarina.

Officially the world's fastest road car, currently with the largest battery pack of any production road car globally, its 120kw power bank feeds the huge 4-wheel drive 1,900bhp powertrain to reach a maximum speed of 217mph (350kph) and a 0-62mph (0-100kph) in under 2 seconds.

Absolutely blistering performance, at an astrological price. The inside is as space age and high tech as you could ever imagine and the body design peerless. But could you use this as an everyday car? Of course not, but wouldn't it be fun!

PROS

» World's fastest production road car
» 186mph – 300kmph top speed
» Looks fantastic
» Great design inside and out
» Exclusive

CONS

» Enormously expensive
» Huge estimated depreciation
» Mediocre range considering price
» 2-seater 2 door only
» Not an everyday car

PRICE £2million (Est) $2.5million USD
ON SALE Now

POLESTAR 2

Australia, Canada, Europe, New Zealand and USA

MANUFACTURERS RANGE

200 miles (est.) – 405km

HOME CHARGE TIME

5 hours (est.)

RAPID CHARGE TIME

26 minutes to 80%

The Polestar luxury brand is an offshoot of Volvo and both owned by China's Geely company, also owners of Lotus, Proton and Manganese holdings - the maker of the London Black Cabs in electric hybrid form.

The Polestar 2 is a true BEV and its design is both retro and futuristic, especially its suave but functional interior.

Its large 75kW battery pack feeds a 402bhp all-wheel drive electric propulsion system that is capable of powering the Polestar 2 from 0-62mph (100kph) in just 4.7s.

The cars batteries are capable of being charged by the very latest 150kW rapid chargers, from 10% to 80% in just 26 minutes. Its top speed is 155mph (250kph).

PROS

» Spacious 4 seat, 4 door family saloon
» Looks great with a twist of retro
» Prestigious badge

CONS

» Expensive
» Pricier than petrol version
» Less range than its peers

PRICE £52,000 (Est) - $60,000 US
ON SALE 2020

PORSCHE TAYCAN

Australia, Canada, Europe, New Zealand and USA

MANUFACTURERS RANGE
220 miles (est.) – 355 km

HOME CHARGE TIME
6 hours (est.)

RAPID CHARGE TIME
30 minutes to 80% (est.)

The Taycan is Porsche's first ever full BEV. The Taycan is also one of Porsche's most anticipated new models ever. The design is pure Porsche with a twist, it's almost as though Ferraris chief designer has been doing some undercover freelance work for Porsche, such are the Taycan's beautiful flowing lines.

But all this beauty and technology does come at a price. The range is estimated to start at £90,000 GBP ($117,000 USD). Nevertheless, compared to similarly specified Aston Martin's 4 door, 4 seat sports coupe – the Rapide-e, the Porsche appears to be quite a bargain.

Rated at a huge 590bhp, its motor delivers 0-62mph (100kph) in just 3.s and with its famed reputation for superb engineering, fit and finish, the Taycan is bound to disrupt the established luxury brands at this pricing level.

PROS

» 4 seat, 4 door sports coupe
» Looks fantastic
» Prestigious badge
» Great performance

CONS

» Very expensive
» Pricier than rivals
» Less range than cheaper rivals

PRICE £90,000 (Est) - $117,000 USD
ON SALE 2020

SEAT EL-BORN

Australia, Canada, Europe, New Zealand and USA

MANUFACTURERS RANGE
250 miles (est.) – 402km

HOME CHARGE TIME
6 hours (est.)

RAPID CHARGE TIME
47 minutes to 80% (est.)

The EL-Born is specifically Seat's version of the new Volkswagen iD hatch, based on the groups new MEB platform.

The EL-Born boasts a range of 250 miles (402 km) and can accelerate to 62mph (100kmh) in just 7.5 seconds.

Using a 100kW ultra charger, the EL-Born can be charged to 80% in just 47 minutes, or 6 hours using a domestic type 2 32A charge box.

The interior of the EL-Born is setting the groups standards for the next 5 years in internet connectivity and entertainment systems. There is Level 2 autonomous capability too, when the time is right.

This new electric Seat will be competing directly with cars such as the Nissan Leaf, and new generation Renault Zoe.

PROS
- Spacious 5 seat, 4 door small SUV
- Looks great
- Good range

CONS
- Expensive
- Pricier than petrol equivalent
- Less range than cheaper rivals

PRICE £30,000 (Est) - $38,0000 USD
ON SALE 2020

SKODA Vision E

Australia, Canada, Europe, New Zealand and USA

MANUFACTURERS RANGE
290 miles (est.) - 377 km

HOME CHARGE TIME
5 hours (est.)

RAPID CHARGE TIME
30 minutes to 80%

Unveiled in 2017, Skoda's Vision-E is the first BEV offered by the company. This sports SUV will be driven by two electric motors producing 302 bhp to all 4 wheels, making it a very practical proposition and alternative to the current crop of ICE SUV's.

Skoda Vision-E draws on the experience of its parent company, the VW group and will use the same propulsion system that they use in similar sized cars across their EV ranges, including Seat and VW. The Vison-E is packed with a host of useful technology, including the latest level 3 autonomous driving capability.

Its 303bhp drivetrain powers all 4 wheels and is capable of a range of 290m (467km) on a full charge. Rapid charging the Vision-E to full can be achieved to 80% in just 30 minutes using one of the latest 100kW ultra chargers. Its top speed is a rather brisk 6.8s 0-62mph (100kph). This flexible design may well be worth the wait.

PROS

» Futuristic 4 seat, 4 door coupe
» Looks superb
» Quality interior

CONS

» Unknown performance
» Pricier than petrol equivalents

PRICE £30,000 (Est) – $37,000 USD
ON SALE 2020

TESLA Roadster

Australia, Canada, Europe, New Zealand and USA

MANUFACTURERS RANGE
400 miles (est.) – 643 km

HOME CHARGE TIME
5 hours (est.)

RAPID CHARGE TIME
30 minutes to 80%

At launch, Elon Musk made a number of bold statements about the Roadster's performance. He claimed a top speed of 250mph (402kph), quarter of a mile from standing start in just 8.8s and a 0-62mph (100kph) time of just 2.0s. That was then and this is now. Of course, we expect all manufacturers to overblow specifications to garner interest in the early days of prototyping, so only time will tell.

One thing is for sure, that with a battery pack of 200kW, the estimated range of 400 miles (644km) is eminently achievable.

The Tesla Roadster will use a dual motor set up. One at the back and one at the front, driving all 4 wheels. Although the roadster comes with four seats, this is strictly a 2+2 design. If the hype matches the specification, this will be a very special, if not hugely expensive, electric sports car.

PROS

» 2 seat, 2 door Super car
» Looks fantastic
» Incredible performance

CONS

» Hugely expensive
» Pricier than petrol sports coupes
» Build quality not as good as competitors

PRICE £170,000 (Est) - $200,000 USD
ON SALE 2020

TESLA Model Y

Australia, Canada, Europe, New Zealand and USA

MANUFACTURERS RANGE

200 miles (est.) - 322 km

HOME CHARGE TIME

5 hours (est.)

RAPID CHARGE TIME

30 minutes to 80% (est.)

In standard mode, Tesla's new Model Y is an impressively specified compact SUV. This model is forecast to eventually take the majority of sales outside the USA. SUVs are more popular than saloon/sedan car designs and the Model Y is set to fill a void in current BEV worldwide production and launches.

Although the Model Y is based on Tesla's Model 3, it offers a far more practical and flexible solution for the likes of young families and adventurers who on a whim, want to throw everything in the back, hang their bikes on the back and escape to the coast or into the country for a long weekend, knowing that you have a vehicle that is very capable, practical and great fun to drive.

Powered at entry level by a 201bhp drivetrain, the Model Y is capable of accelerating from 0-62mph (100kph) in just 6.2 seconds. It can do all this, while providing a top speed of 125mph (201kph) and a range of 200miles (322km) on full charge.

PROS

» Compact 4 seat, 5 door SUV
» Looks great
» Known reliability

CONS

» Pricier than petrol rivals
» Less range than cheaper rivals

PRICE £35,000 (Est) - $39,000 USD
ON SALE 2020

VAUXHALL/OPEL CORSA-e

Australia, Canada, Europe, New Zealand

MANUFACTURERS RANGE
200 miles – 321 km

HOME CHARGE TIME
5 hours (est.)

RAPID CHARGE TIME
40 minutes to 80% (est.)

The new Vauxhall/Opel Corsa-e is planned to launch the company's path to electrification. Based on the next generation of its ICE siblings, but with a completely redesigned floor pan to accommodate the Corsa-e's 50kWh battery pack that can push the cars 134bhp motor from 0-62mph (100kph) in just 8.6s.

Using the latest rapid charger network, you can top up the Corsa-e's battery from 10% to 80% in 40 minutes. Using a 32A home charge box will take roughly 5 hours for the same charge.

The car itself is only available in 4-door form and will comfortably seat 4 passengers although it's classed as a 5-seat car. The Corsa-e's range is estimated to be 200 miles (321km) making it a practical replacement for most small family hatchbacks. Priced competitively for an EV, it is destined to be a success for the Vauxhall/Opel group.

PROS

» Small 5 seat, 5 door-compact hatch
» Looks good
» Flexible load space

CONS

» Pricier than petrol version
» Less range than rivals

PRICE £27,000 (Est)
ON SALE 2020

VOLKSWAGEN iD

Australia, Canada, Europe, New Zealand and USA

MANUFACTURERS RANGE
250 miles – 402 km

HOME CHARGE TIME
6 hours (est.)

RAPID CHARGE TIME
25 minutes to 80% (est.)

The Volkswagen iD is set to be the brands first all-electric production car based on the company's EV only MEB-platform. The VW iD will be available with 3 battery options: standard-range, mid-range and long-range, using the same power train in all three variants. The iD will be in direct competition with BMW's i3 and Renault Zoe and offer similar specification and performance.

The iDs powertrain will be fed by a 60kW battery pack that can be rapid charged in 40 minutes to 80% using the new 100kW rapid charge network. Acceleration is 0-62mph (100kph) in 8s and it is capable of reaching a top speed of 100mph (161kph).

VW are at pains to make it clear that the arrival of their new iD heralds a third coming for the brand. First it was the Beetle, then the Golf, now it's the iD and other siblings to come. If you want a sensibly priced city EV, this could be the one.

PROS

» Compact 5 seat, 5 door-hatch
» Looks superb
» Prestigious badge

CONS

» Expensive
» Pricier than petrol rivals
» Less range than cheaper rivals

PRICE £32,000 (Est) - $38,000 USD
ON SALE 2020

Price – the holy grail

With price EV v ICE parity not forecast until 2024 at the earliest, do you wait, or do you buy now? This decision will be guided by whether you plan to buy a used or new EV. Nevertheless, if you plan to buy used, don't write off the idea of buying a new EV on cost alone. Make sure that you check out new car PCP deals (Personal Contract Plan) or PRP deals (Personal Rental Plan) first. New car deals often require just one month's payment as a deposit and then regular monthly payments can work out the same, if not lower than buying a used car with HP (Hire Purchase) or bank loan payments over three years. The downside is that you don't own the car at the end of a PCP or PRP, but you should normally end up with equity, to put down a deposit again for your next new car for the next three or four years. So why hang on to a depreciating chunk of metal?

With the UK, the first major economy in the world to commit to zero carbon emissions by 2050 and the banning of petrol and diesel vehicles by 2040, we will all eventually have no choice but to own and drive electric cars. But while these legal targets are still more than two decades away, future evolution is proceeding at pace, according to the most recent government data from the EU, UK and USA.

With sales of electric vehicles revealing their highest share of the vehicle market on record year on year, the future is certainly looking green, quiet and pollution free.

FACT

EV's are more efficient than ICE's. An EV is on average 60% more efficient than an ICE.

THREE FACTS THAT MAY HELP IN YOUR DECISION-MAKING PROCESS:

1. Taking into account vehicle price, fuel, Insurance, depreciation, servicing and tax, several experts have revealed that EV's are cheaper to own than ICE vehicles in the UK, Mainland Europe, Japan and the US.
2. Government subsidies and grants are making EV's cheaper in most countries and of course, EV's attract no penalty when entering Zero Emission zones.
3. Some charge points in public places are free to use. Where could you do that legally with petrol/gas or diesel?

Figure 20 **Car Showroom - BMW i3**

10. OWNERSHIP

Since entering the green technology sector in 2000 I have been the owner of several LEV's (light electric vehicles) including several electric bicycles. My one desire was to get hold of an electric car. But the issues were always, when to buy and is the technology ready yet. Is the range good enough? How reliable are they? Most of us ask exactly the same questions in any new technology purchase, whether it be TV's, smart speakers or mobile phones. There will always be the promise of better technology on the horizon. So, exactly when do you bow to fate?

I still had no intention of owning an electric car until the range increased and the charging infrastructure got better. But that changed at the end of a meeting with a friend and business colleague at Harwell research centre in Didcot in 2016. Mike had been a first adopter of EV's way back in 2006, when he purchased and drove his first G Wizz electric car. Since then he has run the world's largest electric vehicle owners club and owned many different electric cars over the years, including his Nissan leaf. Our meeting overrun and I needed to get back to Didcot Parkway station to catch my train back home. Mike suggested that he would take me to the station in his car. He unplugged the 'Leaf' from the car park charger and from that point on, it was 'hellfire and brimstone'. I was overwhelmed by the instant torque, road handling and acceleration of this average, rather goofy looking small hatchback. Within minutes we had made it to the station, and I got my train back with seconds to spare. This was one of those game changing moments for me. On the train, I hurriedly trawled the web to find available electric cars in my location and booked three test drives the following weekend. I was truly won over by the sheer exhilaration and experience of electric cars from that point on.

I test drove a Nissan Leaf, an e-Golf and a BMW i3. After testing all three, there was no loser, but I chose on gut instinct and decided that my first electric car purchase was going to be the futuristic BMW i3. The exterior design, you either love or hate, but this was the most revolutionary car of the three. Using a lightweight carbon fibre shell and rear wheel drive, to me it ticked all the boxes and definitely felt the nippiest of all three. I live in a small sleepy Cotswold village. Unlike London, where EVs are almost as common as bicycles in Beijing, there are none where I live or anywhere close to where I live, for that matter. I was certain that I was the only EV owner within a 5-mile radius of home. That means most of my friends and neighbours knew little about EV's and thus I became an ambassador of the green, quiet, EV lifestyle.

Of course, not all people were positive about my new EV. The most frequent objection I heard was that 'the car is ugly' or 'it doesn't have enough range for my liking'. But I found myself constantly explaining that it is perfectly fine for 90% of my driving requirements and I normally go two or three days without recharging. The cynics then begin to see how an electric car could meet their needs too. In most cases, after a few minutes of discussion, you could sense a eureka moment developing. In the early days of EV ownership, I felt that I'd been converted to a new religion and transformed into an EV evangelist against my better judgement!

It was the first time for years that I would randomly pick up the car keys and go out for a drive, just for the hell of it. I'm sure that you will feel this way too. Driving the car was an absolute delight. It was quiet with great power reserves and a comfortable, but often hard ride on uneven roads, but typical of BMW suspension tuning. With every drive beyond my local zone, my confidence grew. My initial bouts of range anxiety faded to a distant memory. I ended up taking the car on far longer trips than the stated range, ensuring that I knew where the rapid charging stations were on the way. In the end, I had no problem traveling to London Gatwick airport and back – a round trip of more than 250 miles (402 km), and this was in a car with an average range of 85 miles maximum. Since that initial purchase, BMW i3s and Nissan Leaf ranges have both increased dramatically by more than 50 percent.

To ensure that most of my charging was at home, more for convenience, I installed a 32A Type 2 charging point in my garage, although these are now heavily subsidised in the UK and most other countries too. The i3 could also be plugged directly into a standard domestic socket, but it took up to 8 hours to top up the battery. Using the 32A Type 2 socket meant I could charge it fully in about 3 hours from flat. Over the first year of running the car, I did notice a significant reduction in range between the winter and summer running. Ambient temperature can really affect range on all electric cars, so be aware that the stated range of an EV is almost always based on summer temperature. You will need to subtract roughly 25% for winter range.

Having the first electric car in the village was akin to getting the first rooftop solar system installed. It helps stimulate interest and discussion about new technology and how it could fit into peoples own lifestyle. Simply driving it locally helped break down the barriers that prevented my neighbours considering buying their own electric car. I am happy to say that there are now at least 6 electric cars locally and hopefully more to come. Nevertheless, the initial shock of owning an electric car is discovering when you turn the car on. It is completely silent. Electric cars only emit the feintest hum, even on the road.

The range anxiety that I initially experienced became a thing of the past. With models like the Tesla P100D and the KIA e-Niro boasting real world ranges of more than 300 miles (480km) on a full battery, range on future EV's will only get better, as the cost of batteries comes down year on year. In 2017, a million EVs were sold globally and this is forecast to double over the next two years. China has even announced its goal of having more than a million EV's on its roads by 2020.

Of course, EV ownership is not completely perfect; nothing ever is. EV's remain expensive, but prices are starting to fall dramatically. A used 3-year old Nissan Leaf will cost in the region of £14,000 GBP or $18,000 USD. Whilst a 4-year old Tesla Model S can be bought for around £35,000 GBP or $45,000 USD.

EV batteries have been found to outperform all forecasts, well after 100,000 miles (161,000km) of use, suggesting that EV residuals will maintain strong value for many years to come.

Even when you reach 100,000 miles in your EV, many car companies plan to allow customers to swap their old battery packs and recycle them to be used in BESS (Battery Energy Storage Systems). There are also some specialist recycling companies that are already recovering old EV packs to store solar and wind power for use at peak generating times to feed back power in to the electricity grid.

It's amazing that in the space of about 4 years since I bought my first EV, the novelty of seeing EV's in towns and car parks has swiftly eroded. There will soon become a time when modern families never visit a traditional petrol station and these gas stations will hopefully go the way of the dodo. Why pay at least four times the cost for petrol or diesel (rising often on a weekly basis), when for about a quarter of the price, you can fill up with cheaper electricity and run a faster, more exciting, zero tailpipe polluting green car? For me and hopefully you, the future is now.

FACT

EV's have low maintenance and running costs due to very few moving parts.

11. THE FUTURE

Electric cars are vitally important to the future of the car industry and to the ecosystem that we all rely on. Though, it is still not certain what form the electric car will eventually take, and whether the general driving public will give their wholehearted approval. One thing that is certain, is that the depletion of oil supplies, fears of noise and air pollution and the difficulties of recycling petrol and diesel-powered cars, are all motives that seem to be powering toward the realisation and success of the electric car.

Of course, the worlds dependence on oil and its inevitable demise was always certain to generate serious research and innovation into alternative fuel and transport. But now there is an even more compelling reason to dump the oil burner for a cleaner greener alternative.

Is the future actually hydrogen? In my opinion no, and the reason I say this is that extracting hydrogen actually uses more than three times the energy than it actually produces, so it's not exactly green at the moment. Additionally, hydrogen fuel cells produce an invariable level output of power, akin to a trickle charger. This means that it requires a battery pack to feed and run the variable speed electric power train. So, a hydrogen powered car is effectively a battery powered car, with the additional cost and weight of a fuel cell and hydrogen storage tank. Then there is the safety aspect and hugely expensive hydrogen fuel service station infrastructure that would need to be built.

On the other hand, the transition to an all-electric future is gaining momentum globally. Of course, it will require sizeable investments in research and development to accelerate the deployment of vehicle charging infrastructure and potential legislation for the installation of charging boxes in new and existing homes.

Nevertheless, while electric cars, buses and lorries may account for a miniscule percentage of vehicles on the road today (according to Forbes [27] - 4 million electric vehicles versus over one billion petrol and diesel cars), adoption is accelerating, fast.

Figure 21 **Is This The Future Form of Electric Cars?**

To put things into perspective, C. McKerracher & J. Wu [28], noted that it took more than 20 years to sell the first million electric cars. Now, a million electric vehicles are sold in just four to five months and who is brave enough to forecast what form new EV's will look like in the next 30 to 40 years. One thing is certain – the future is definitely electric (figure 21).

12. TABLES AND DIAGRAMS

Figure **Page**

Figure		Page

13. ACRONYMS & ABBREVIATIONS

3D	Three Dimensional
AC	Alternating current
BEV	Battery Electric Vehicle
BESS	Battery Energy Storage System
CCS	Combined Charging System (European standard)
CHAdeMO	Charge on the Move (Japanese standard)
DC	Direct Current
DC/DC	Direct Current to Direct Current converter
DOT	Department of Transport
DfT	Department for Transport
EPA	Environmental Protection Agency
EV	Electric Vehicle
FCEV	Fuel Cell Electric Vehicle
GBP	Great Britain Pound
GB/T	GuoBiao/Tuiijian (National Standard/Recommended (Chinese))
HEV	Hybrid Electric Vehicle
HP	Hire Purchase
ICE	Internal Combustion Engine
kW	Kilo Watt

LFP	Lithium Iron Phosphate (battery)
Li-ion	Lithium ion (battery)
LMO	Lithium Manganese Spinel (battery)
LV	Light Vehicle or Low Voltage
MPC	Miles Per Charge
NCA	Lithium Nickel Cobalt Aluminium (battery)
NCAP	New Car Assessment Program
NMC	Lithium Nickel Manganese Cobalt (battery)
NEDC	New European Driving Cycle
OEM	Original Equipment Manufacturer
PCP	Personal Contract Plan
PEV	Pure Electric Vehicle
PHEV	Plug in Hybrid Electric Vehicle
PRP	Personal Rental Plan
R&D	Research and Development
REHEV	Range Extending Hybrid Electric Vehicle
SUV	Sports Utility Vehicle
USD	United States Dollar
US DOT	United States Department of Transport
WHLV	World Harmonised Light Vehicle test.

14. IMAGE COPYRIGHT ACKNOWLEDGEMENTS AND EV INDEX

Image	Page

15. REFERENCES

[1] Department of Energy (US 2018). The history of the electric vehicle.

[2] Department for Transport (UK 2015). The green car revolution across US cities.

[3] International Energy Agency. Global EV outlook 2018. (March 2019).

[4] International Economic Development Council (2013). Creating the clean energy economy.

[5] Marc Walberger et al. Tesla – a fortune retrieved. (2017)

[6] European Environment Agency. Electric Vehicles In Europe. (2016)

[7] Forbes Journal. Chinas Electric Vehicle Leaders. Who are they? (2019)

[8] London Victoria General Insurance Survey. (May 2019). Busting the myths of the electric car. (March 2019)

[9] Department for Transport (UK). National travel survey. (2016)

[10] Deloitte. New Market. New Entrants. New Challenges. Battery Electric Vehicles. (2018).

[11] PWC. Charging Ahead. The need to upscale UK electric vehicle charging infrastructure. (April 2018)

[12] Union of Concerned Scientists (USA). Electric Vehicles. (2018)

[13] US Securities and Exchange Commission. Electric Vehicle Federal Law. (2018)

[14] Navigant research 2019: Global Forecasts for Light Duty Plug-In Hybrid and Battery EV Sales and Populations: 2018-2030.

[15] Boston Consulting Group 2018. The future of battery production for electric vehicles

[16] Yamindhar Reddy Bhavanam. Combined Power System Planning Policy Proposition For Future Electric Vehicle Charging Infrastructure. (2015)

[17] Nissan Statutory Propriety Data Release. (2015)

[18] Reisch, Marc S. Solid state batteries inch their way towards commercialization. (2017)

[19] Qi, Zhaoxiang, et al. Journal of Vacuum Science & Technology B, Nanotechnology and Microelectronics: Materials, Processing, Measurement, and Phenomena. (July 2017).

[20] Li, Zhilin; Chen, Lianlian; Meng, Sheng; Guo, Liwei; Huang, Jiao; Liu, Yu; Wang, Wenjun; Chen, Xiaolong. Field and temperature dependence of intrinsic diamagnetism in graphene: Theory and experiment. Phys. Rev. B. (2015).

[21] Dyson press release. New Electric Vehicle Design Facility In UK. (2018).

[22] Machine Design (USA). Permanent Magnet Technology. (2019)

[23] Aoife Foley, et al. Electric Vehicles and Energy Storage – A case study on Ireland (2016)

[24] McKinsey. Electric Vehicle Index. (2018)

[25] McKinsey. Automotive and Production report. (2017)

[26] Deloitte. New Market. New Entrants. New Challenges. Battery Electric Vehicles. (2018).

[27] Forbes. When European Fuel Efficiency Rules Bite, Buyers can choose the Electric Car or Bus (June 2019)

[28] C. McKerracher & J. Wu. Mobility Transition (2018)

Printed in Great Britain
by Amazon